BrightRED Study Guide

Curriculum for Excellence

N4

BIOLOGY

Margaret Cook and Fred Thornhill

First published in 2015 by:
Bright Red Publishing Ltd
1 Torphichen Street
Edinburgh
EH3 8HX

Reprinted with corrections in 2017

Copyright © Bright Red Publishing Ltd 2015

Cover image © Caleb Rutherford

All rights reserved. No part of this publication may be reproduced, stored in a retrieval system, or transmitted in any form or by any means, electronic, mechanical, photocopying, recording or otherwise, without prior permission in writing from the publisher.

The rights of Margaret Cook and Fred Thornhill to be identified as the authors of this work have been asserted by them in accordance with sections 77 and 78 of the Copyright, Designs and Patents Act 1988.

A CIP record for this book is available from the British Library

ISBN 978-1-906736-46-0

With thanks to:
PDQ Digital Media Solutions Ltd, Bungay (layout and illustrations), Ailsa Morrison (copy-edit)

Cover design and series book design by Caleb Rutherford – e i d e t i c

Acknowledgements
Every effort has been made to seek all copyright holders. If any have been overlooked, then Bright Red Publishing will be delighted to make the necessary arrangements.

Permission has been sought from all relevant copyright holders and Bright Red Publishing are grateful for the use of the following:
Krzysztof Szkurlatowski/Freeimages.com (p 6); Mulletsrokk (CC BY-SA 3.0)[1] (p 6); Kristian Peters (CC BY-SA 3.0)[1] (p 6); Jacob Davies (CC BY-SA 2.0)[2] (p 8); Joshua Newman Design (p 8); Flavio Takemoto/Freeimages.com (p 10); 123dan321/Freeimages.com (p 15); lumix2004/freeimages.com (p 16); lumix2004/freeimages.com (p 16); Jordi Play (CC BY-SA 2.0)[2] (p 16); Alan Levine (CC BY-SA 2.0)[2] (p 20); US Department of Agriculture (CC BY 2.0)[3] (p 21); Col Ford and Natasha de Vere (CC BY 2.0)[3] (p 22); Todd Nienkerk (CC BY-SA 2.0)[2] (p 23); Jeff Alworth (CC BY 2.0)[3] (p 23); Jeffrey Keeton (CC BY 2.0)[3] (p 23); Jeffrey Keeton (CC BY 2.0)[3] (p 23); Eric Hwang (CC BY-ND 2.0)[4] (p 23); Johann Jaritz (CC BY-SA 3.0)[1] (p 24); Schematic view of the biofuel conversion process © US Department Of Energy, 2010 (p 25); Vasant Dave/Freeimages.com (p 27); Enjoylife2/iStock.com (p 36); Poster © Foundation for Biomedical Research (p 37); Poster © Choose Cruelty Free (choosecrueltyfree.org.au) (p 37); Poster © Choose Cruelty Free (choosecrueltyfree.org.au) (p 37); Image from 'The Pathogen Profile Dictionary' © Journal of Undergraduate Biological Studies (p 42); 500paolo (CC BY 3.0)[6] (p 42); Merje Toome (CC BY-SA 3.0)[1] (p 43); Public Domain (p 43); Kalyanvarma (CC BY-SA 3.0)[1] (p 43); John Liu (CC BY 2.0)[3] (p 44); Andrew (CC BY 2.0)[3] (p 44); Andrew (CC BY 2.0)[3] (p 44); Ian Muttoo (CC BY-SA 2.0)[2] (p 45); S. Rae (CC BY 2.0)[3] (p 45); Tony Hisgett (CC BY 2.0)[3] (p 45); freefoodphotos.com (CC BY 3.0)[6] (p 45); Armin Kübelbeck (CC BY-SA 3.0)[1] (p 47); Joi Ito (CC BY 2.0)[3] (p 50); Jesuino Souza/Freeimages.com (p 51); Jesuino Souza/Freeimages.com (p 51); Maja Petric/Freeimages.com (p 51); Olivier Le Moal/ Dreamstime.com (p 52); Ralph Reski (CC BY-SA 3.0)[1] (p 53); Gita Kulinica/Dreamstime.com (p 53); Raidarmax (CC BY-SA 3.0)[1] (p 54); James Ball (CC BY 2.0)[3] (p 54); Gordon (CC BY-SA 2.0)[2] (p 54); Gordon (CC BY-SA 2.0)[2] (p 54); Catrin Austin (CC BY 2.0)[3] (p 54); Tatjana Romanova/Shutterstock.com (p 54); Gang Liu/Shutterstock.com (p 54); Don Kosmayer/123rf (pp 54 & 55); nrt/Shutterstock.com (pp 54 & 55); Sam Howzit (CC BY 2.0)[3] (p 56); Carissa Rogers (CC BY 2.0)[3] (p 56); Carissa Rogers (CC BY 2.0)[3] (p 56); CWA Studios/Shutterstock.com (p 62); Sebastian Kaulitzki/Shutterstock.com (p 62); Praisaeng/Shutterstock.com (p 64); Trevor Leyenhorst (CC BY 2.0)[3] (p 64); Ian Rentoul/Shutterstock.com (p 69); 2 images by Kev Chapman (CC BY 2.0)[3] (p 69); Doug Beckers (CC BY-SA 2.0)[2] (p 70); Poster © GB Non-native species secretariat 2015 (Photo by Michael Grabowski & drawings by Mike Dobson/FBA) (p 71); rodehi/iStock (p 71); Greenpeace China (CC BY 2.0)[3] (p 72); Jesse Allen and Robert Simmon/NASA Earth Observatory (public domain) (p 72); Bundesarchiv, Bild 183-48195-0006 (CC BY-SA 3.0 DE)[5] (p 72); Josh Estey/AusAID (CC BY 2.0)[3] (p 74); Josh Estey/AusAID (CC BY 2.0)[3] (p 74); American Spirit/Shutterstock.com (p 74); Frits Ahlefeldt-Laurvig (CC BY-ND 2.0)[4] (p 74); USFS Gila National Forest (CC BY-SA 2.0)[2] (p 76); Katkosia (CC BY-SA 3.0)[1] (p 76); miniwiki.org (public domain) (p 76); David Rydevik (public domain) (p 77); U.S. Navy photo by Photographer's Mate 2nd Class Philip A. McDaniel (public domain) (p 77); Hansueli Krapf (CC BY-SA 3.0)[1] (p 77); Harry Rose (CC BY 2.0)[3] (p 79); Carl Lewis (CC BY 2.0)[3] (p 80); Rasbak (CC BY-SA 3.0)[1] (p 80); Image taken from 'The Project Gutenberg EBook of The Tomato,' by Paul Work (public domain) (p 81); Nousha/iStock.com (p 81); karenandkerry (CC BY 2.0)[3] (p 83); karenandkerry (CC BY 2.0)[3] (p 83); U.S. National Archives and Records Administration (public domain) (p 83); wk1003mike/Shutterstock.com (p 83); Fenrisulfir (CC BY-SA 3.0)[1] (p 84); davebloggs007 (CC BY 2.0)[3] (p 85); Alexandr Trubetskoy (CC BY-SA 3.0)[1] (p 85); Dr. Jennifer L. Graham/U.S. Geological Survey (CC BY 2.0)[3] (p 85); Tambako The Jaguar (CC BY-ND 2.0)[4] (p 86); Neil McIntosh (CC BY 2.0)[3] (p 87); Jim, the Photographer (CC BY 2.0)[3] (p 87); dobrotica/iStock (p 87); dorena-wm (CC BY-ND 2.0)[4] (p 87); Arturo de Frias Marques (CC BY-SA 4.0)[7] (p 88); Ted (CC BY-SA 2.0)[2] (p 88); Lisa Leonardelli (CC BY-SA 2.0)[2] (p 88); Martin Konopacki (CC BY-SA 2.0)[2] (p 88); Katie Steiger-Meister/USFWS (p 89); Bevis Chin (CC BY-ND 2.0)[4] (p 89); Tam Warner Minton, TravelsWithTam.com (p 89); Beth Swanson/Shutterstock.com (p 89); Steven C Wilson (CC BY-ND 2.0)[4] (p 90); Neil McIntosh (CC BY 2.0)[3] (p 90); Neil McIntosh (CC BY 2.0)[3] (p 90); Liam Quinn (CC BY-SA 2.0)[2] (p 90); Eric Inafuku (CC BY 2.0)[3] (p 91); Kim Briers/Shutterstock.com (p 91); Yblieb (CC BY-SA 3.0)[1] (p 92); Donna Drewhurst/U. S. Fish and Wildlife Service (public domain) (p 92); Bright Red Publishing would also like to thank the Scottish Qualifications Authority for use of Past Exam Questions: SQA Standard Grade General Biology 2012 Paper. Answers do not emanate from the SQA (p 95).

(CC BY-SA 3.0)[1]	http://creativecommons.org/licenses/by-sa/3.0/
(CC BY-SA 2.0)[2]	http://creativecommons.org/licenses/by-sa/2.0/
(CC BY 2.0)[3]	http://creativecommons.org/licenses/by/2.0/
(CC BY-ND 2.0)[4]	http://creativecommons.org/licenses/by-nd/2.0/
(CC BY-SA 3.0 DE)[5]	http://creativecommons.org/licenses/by-sa/3.0/de/
(CC BY 3.0)[6]	http://creativecommons.org/licenses/by/3.0/
(CC BY-SA 4.0)[7]	http://creativecommons.org/licenses/by-sa/4.0/

Printed in the UK

CONTENTS

BRIGHTRED STUDY GUIDE – NATIONAL 4 BIOLOGY

Introducing National 4 Biology 4

1 CELL BIOLOGY

Cell division and its role in growth and repair
Cell structure & The functions of cell parts 6
Cell division 8

DNA, genes and chromosomes
Genetic information & How genes work 10
Passing on the genetic information 12

Therapeutic uses of cells
Genetic engineering 14
Stem cells 16

Properties of enzymes and their use in industries
Properties of enzymes & Enzymes at work 18
The use of enzymes in industry 20

Properties of microorganisms and their use in industries
Useful characteristics of microorganisms & Traditional uses of yeast 22
Traditional uses of bacteria & Other uses of microorganisms 24

Photosynthesis
The process of photosynthesis & The factors needed for photosynthesis 26
Measuring the rate & Limiting factors 28

Respiration
The process of respiration & Respiration with and without oxygen 30
Factors affecting the rate of respiration 32

Controversial biological procedures
Gene therapy, pharming and transgenic organisms .. 34
Cloning and animal research 36

2 MULTICELLULAR ORGANISMS

Sexual and asexual reproduction
Differences between and the advantages of sexual and asexual reproduction 38
Production of sex cells in flowering plants and mammals 40
Success rates of sexual reproduction & Methods of asexual reproduction 42

Propagating and growing plants
Sexual and asexual reproduction 44
Asexual propagation: natural methods 46
Asexual propagation: artificial methods 48

Commercial uses of plants
Food & Fuels 50
Medicines 52

Genetic information
Species & Variation 54
Inheritance 56

Growth and development in different organisms
Growth curves, Growth in plants & Growth in animals 58
Factors needed for growth in plants and animals ... 60
The effect of radiation and chemicals on growth ... 62

Maintaining stable body conditions
The nervous system 64
Homeostasis & Maintaining body temperature and blood glucose levels 66

3 LIFE ON EARTH

Animal and plant species depend on each other
Biomes, Ecosystems & Biotic factors affecting organisms 68
Sampling techniques & The effect of adding or removing species on other species 70

Impact of population growth and natural hazards on biodiversity
Human population growth & Ecological footprints ... 72
Habitat destruction & Other effects of human activity on biodiversity 74
Impact of natural hazards on biodiversity 76

The nitrogen cycle
Recycling 78
Compost heaps & Water culture experiments 80

Fertiliser design and environmental impact of fertilisers
Mineral deficiencies & Fertiliser design 82
Different fertiliser types & Eutrophication 84

Adaptations for survival
The need for and the types of adaptation 86
Examples of adaptations 88

Learned behaviour in response to stimuli
Innate behaviour 90
Learned behaviour 92

COURSE ASSESSMENT

An overview of assessment 94
Sample questions 96

INTRODUCTION

INTRODUCING NATIONAL 4 BIOLOGY

During this course we hope you will develop and apply skills for learning, skills for life and skills for work.

Biology is the study of living organisms. This makes it relevant to us all. Biology is an increasingly important subject and is involved in the search for solutions to many of the world's problems.

THE NATIONAL 4 BIOLOGY COURSE

National 4 Biology encourages you to become:

- a more confident learner
- a responsible citizen with an informed and ethical view of complex biological issues, through the study of relevant areas such as health, environment and sustainability
- someone who thinks carefully about important issues and is able to make reasoned evaluations.

The National 4 course will help you to develop knowledge and understanding of current biological topics, ranging from the processes which take place inside cells to the relationships between organisms and the environment.

There are four units in the course. The Key Areas are covered in three of the four units.

Key Areas cover cells and cellular processes, leading to an understanding of the industrial and therapeutic uses of cells and microorganisms. Different forms of reproduction are studied, leading to an understanding of the commercial importance of plants. Genetic information, types of life cycles and the importance of controlling internal conditions are also covered. The interdependence of organisms is emphasised together with the ways in which the environment is affected by our activities and by natural hazards. The importance of adaptations and of behavioural responses for survival are also studied.

The Key Areas are:

UNIT 1- CELL BIOLOGY

- cell division and its role in growth and repair
- DNA, genes and chromosomes
- therapeutic use of cells
- properties of enzymes and their use in industries
- properties of microorganisms and their use in industries
- photosynthesis and limiting factors
- factors affecting respiration
- controversial biological procedures

UNIT 2 - MULTICELLULAR ORGANISMS

- sexual and asexual reproduction and their importance for survival of species
- propagating and growing plants
- commercial use of plants
- genetic information
- growth and development of different organisms
- biological actions in response to internal and external changes to maintain stable body conditions

UNIT 3 – LIFE ON EARTH

- dependence of animal and plant species on each other
- impact of population growth and natural hazards on biodiversity
- nitrogen cycle
- fertiliser design and environmental impact of fertilisers
- adaptations for survival
- learned behaviour in response to stimuli linked to species survival

INTERNAL ASSESSMENT

The assessment of the course will be devised and carried out by your school or college and it will cover a range of skills including:

- demonstrating knowledge and understanding of biology by making statements, describing information and providing explanations
- applying biology knowledge to familiar situations, interpreting information and solving problems
- planning and safely carrying out experiments/practical investigations to illustrate effects
- using information handling skills by selecting, presenting and processing information
- making generalisations based on evidence/information
- drawing valid conclusions and giving explanations supported by evidence
- suggesting improvements to experiments/practical investigations
- communicating findings/information

THE ADDED VALUE UNIT

The fourth unit of the course is the Added Value Unit. In this unit you will draw on and apply the skills and knowledge you have learned during the course. The Added Value Unit is assessed through an assignment.

The assignment will be an in-depth study of a topical issue from a key area of the course chosen by you in agreement with your teacher or lecturer. The assignment will be assessed by your teacher or lecturer.

The assignment is carried out under controlled conditions. To prepare for the assessment you will choose and research/investigate an appropriate topic, focusing particularly on the applications and impact on society or the environment, and process the information.

During the assessment you will present evidence of:

- the issue being investigated and its relevance to the environment/society
- a selection of appropriate information from at least two relevant sources
- information displayed appropriately
- a description of the biology of the issue and its impact on the environment/society

HOW WILL THIS GUIDE HELP YOU MEET THE CHALLENGES?

The aim of this book is to help you achieve success in assessment of Key Areas by providing you with a suggested coverage of the Key Areas of the course. Helpful hints are provided throughout the book in the 'Don't forget' features, while there are plenty of opportunities to practise applying your knowledge through 'Things to do and think about'. Some of the skills you will be expected to demonstrate are also covered in the book. These may be in the relevant Key Areas or covered in separate sections.

ONLINE

All of the answers for the questions in this book can be found at www.brightredpublishing.co.uk

UNIT 1: CELL BIOLOGY

CELL DIVISION AND ITS ROLE IN GROWTH AND REPAIR

CELL STRUCTURE & THE FUNCTIONS OF CELL PARTS

CELL STRUCTURE

Cells are the microscopic units from which all living organisms are built. The number of cells differs from organism to organism. Some microscopic organisms consist of just one cell and other multicellular organisms contain many millions of millions of cells. An adult human body is estimated to consist of about 40 million million cells, although estimates vary. The larger an organism, the more cells it will have.

Cells may look very different from each other because some cells are specialised to carry out different functions. However, all cells tend to have the same basic structure and to contain similar parts. It is important to know about the structure of cells in order to understand their important functions.

Listed below are several facts about the structure of typical cells:

- The contents of all cells are surrounded by a cell membrane.
- Inside the cell membrane is a jelly-like substance called cytoplasm. Other structures are present in the cytoplasm.
- Each cell has a nucleus within the cytoplasm.
- Plant cells have a strong cell wall outside of the cell membrane.
- Plant cells have a large fluid-filled space called a vacuole within the cytoplasm.
- Some plant cells have many small green structures called chloroplasts within the cytoplasm.

The following photographs show typical animal and plant cells:

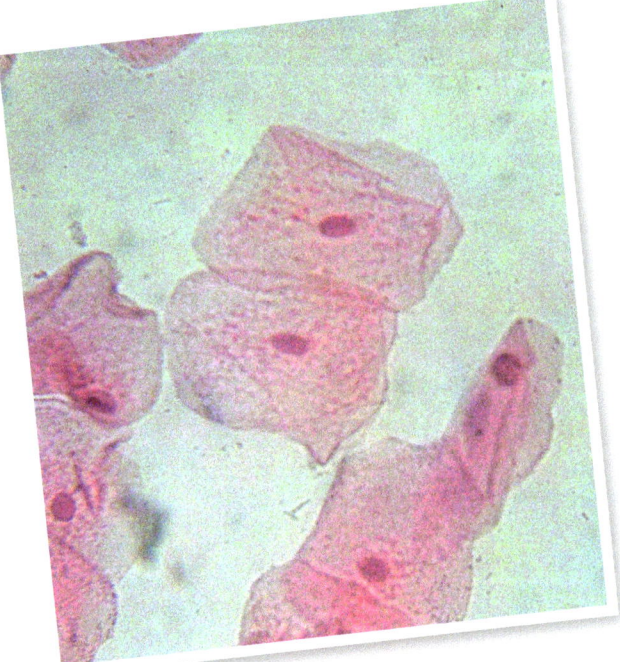
Animal cells

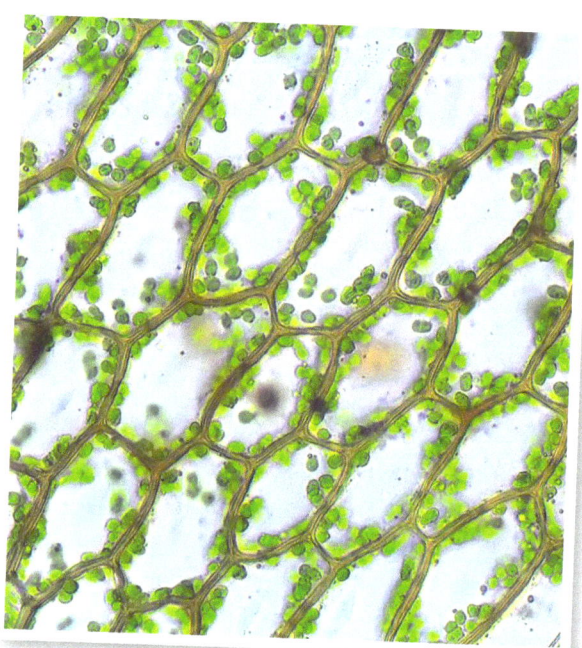
Plant cells

Cell Division and its Role in Growth and Repair – Cell Structure & the Functions of Cell Parts U1

It is not always easy to recognise the various parts of cells and chemicals called stains are often used to make them easier to see when viewed with a microscope.

The diagrams below show the parts of cells more clearly:

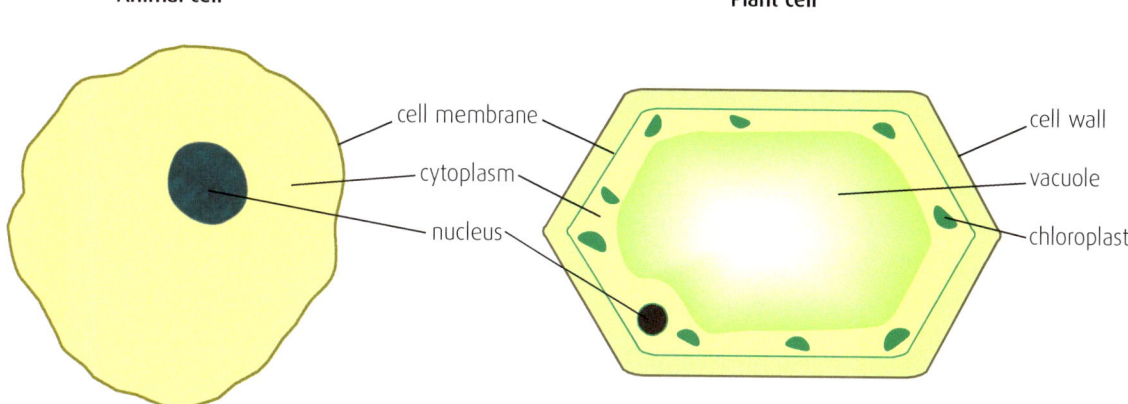

THE FUNCTIONS OF CELL PARTS

Each part of a cell has a function which is important for the cell to survive and continue working:

- The cell membrane controls the movement of substances into and out of the cell.
- The cytoplasm is the location for the essential chemical reactions which take place in cells.
- The nucleus contains the genetic information which controls the cell's development and activities. The genetic information is contained on structures called chromosomes which are found in the nucleus.
- The plant cell wall supports the cell and helps give structure to the plant.
- The vacuole contains a watery solution which keeps the cell firm.
- The chloroplasts contain chlorophyll which absorbs light. The chloroplasts carry out the process of photosynthesis.

SUMMARY
All cells have a cell membrane, cytoplasm and a nucleus. Only plant cells have cell walls and a large vacuole. Some plant cells also contain chloroplasts.

 **DON'T FORGET**

Not all plant cells contain chloroplasts; only those in the green parts of a plant do.

 THINGS TO DO AND THINK ABOUT

1. Copy and complete the following table using information from above.

Name of cell part	Function of cell part	Type of cells containing these parts (Animal only/Plant only/Both)
Cell membrane		
Cell wall		
Chloroplast		
Cytoplasm		
Nucleus		
Vacuole		

2. Name the structures found in the nucleus of cells which contain the genetic information about the cell.
3. Describe the function of the genetic information found in a cell.

CELL DIVISION AND ITS ROLE IN GROWTH AND REPAIR

CELL DIVISION

THE IMPORTANCE OF CELL DIVISION

All new cells are produced from existing cells by the process of cell division.

SINGLE-CELLED ORGANISMS

For single-celled organisms, cell division is the way they normally reproduce. One cell divides into two daughter cells which separate and become new individual organisms.

This happens in bacteria, yeast (a single-celled fungus), amoeba (a single-celled aquatic organism) and many other types of single-celled organisms.

Cell division in amoeba is shown in the diagram below:

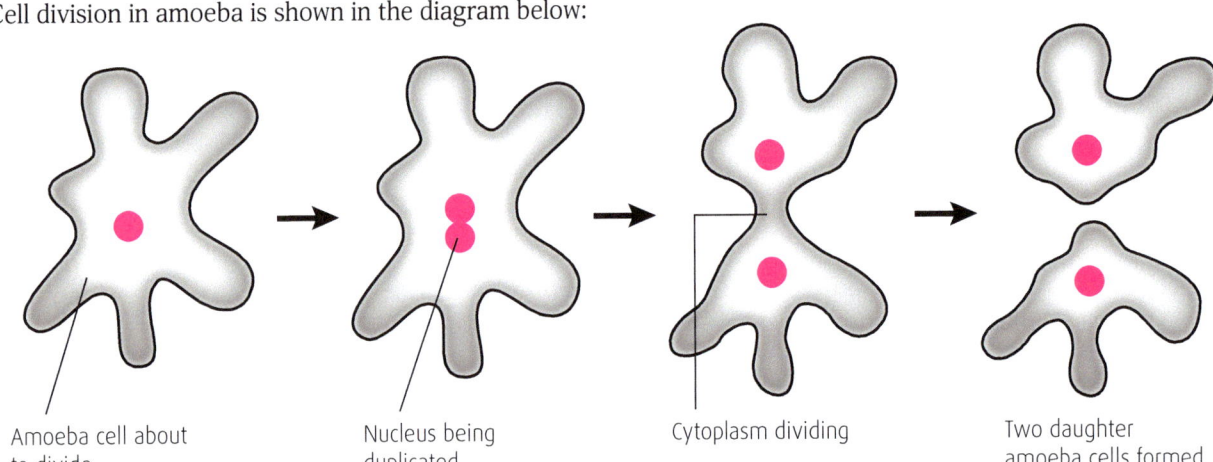

Amoeba cell about to divide | Nucleus being duplicated | Cytoplasm dividing | Two daughter amoeba cells formed

MULTICELLULAR ORGANISMS

For multicellular organisms, cell division increases the number of cells in the organism. This enables an organism to grow and to repair damaged tissues.

It is important that new cells produced by cell division are able to function in the same way as the parent cell. This means that every chromosome in the nucleus of the parent cell must be copied to make a duplicate set of chromosomes. These are then divided between the two daughter cells so that each has exactly the same number of chromosomes as the parent cell.

REGENERATION

Some organisms have great abilities to repair damaged tissue. This is called regeneration.

The ability to regenerate lost tissue depends largely on how many unspecialised cells an organism possesses. The greater the number of unspecialised cells, the greater the regeneration ability.

Starfish can replace damaged or lost arms.

Regeneration in humans is much more restricted. The healing of wounds and broken bones is a limited form of regeneration. Humans can also regenerate damaged liver cells.

Regeneration in starfish

Stages in the regeneration of a leg of a newt (an amphibian)

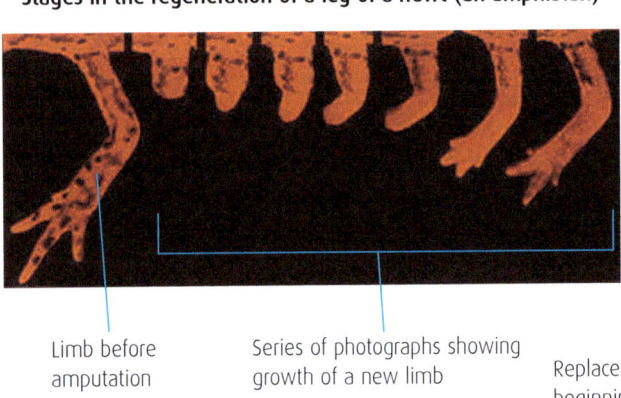

Limb before amputation | Series of photographs showing growth of a new limb | Replacement arm beginning to grow

THE PROCESS OF CELL DIVISION

Chromosomes are long thread-like structures. Normally they are tangled together and are not easily seen. When a cell is about to divide, the chromosomes become shorter and thicker. This makes it possible to see the stages of cell division with a microscope. The stages are shown in the diagram below.

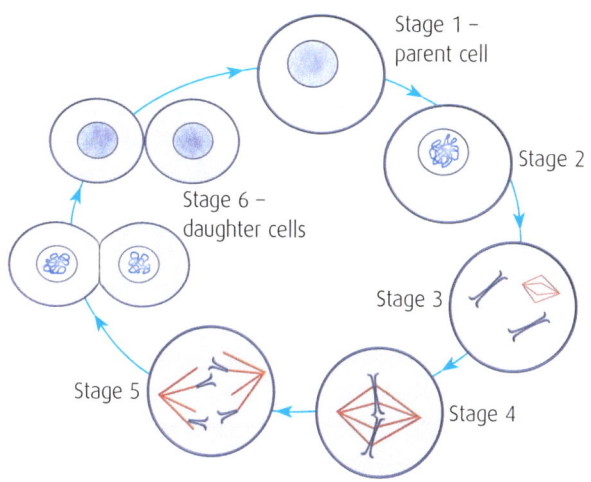

Stage 1 The chromosomes in the parent cell are not visible but they are being duplicated.

Stage 2 The chromosomes become shorter and thicker, making them visible as long thread-like structures.

Stage 3 The chromosomes become free in the cell cytoplasm. Each can be seen attached to its identical duplicate, forming a pair.

Stage 4 The duplicated chromosomes become arranged along the middle of the cell, attached to thin fibres (shown in red) which stretch from one end of the cell to the other.

Stage 5 Each chromosome is separated from its duplicate and the separated strands are pulled to opposite ends of the cell.

Stage 6 The chromosomes become less visible as they form new nuclei at opposite ends of the cell. The cytoplasm divides forming two separate cells.

A PROBLEM WITH CELL DIVISION

Sometimes cell division begins to take place in an uncontrolled manner. This can lead to the formation of a mass of cells called a tumour. A tumour can cause damage to neighbouring cells. This is what happens with cancer, a disease which can be difficult to treat. Treatment normally involves two different methods to try and kill the cancer cells. Powerful drugs can be used. This is called chemotherapy. The other method uses radiation and this is called radiotherapy.

DON'T FORGET

When a cell divides it produces two daughter cells; each daughter cell has the same number of chromosomes as the parent cell.

SUMMARY
- Cell division is a method of reproduction used by single-celled organisms. It is necessary for the growth of multicellular organisms and for the repair of damaged tissue.
- The ability to regenerate damaged tissue depends on the presence of unspecialised cells.
- When a cell divides, two daughter cells are formed. Each daughter cell has the same number of chromosomes as the parent cell.
- Uncontrolled cell division can lead to the formation of cancerous tumours.

THINGS TO DO AND THINK ABOUT

1. What is the purpose of cell division in:
   (a) single-celled (unicellular) organisms?
   (b) multicellular organisms?
2. (a) How many daughter cells are produced when a cell divides?
   (b) How does the number of chromosomes in a daughter cell compare to the number in the parent cell?
3. What process takes place before a cell divides, making sure each daughter cell is able to receive the correct number of chromosomes?
4. (a) Name the disease which may result from uncontrolled cell division.
   (b) Name the two methods of treatment for this disease.

DNA, GENES AND CHROMOSOMES

GENETIC INFORMATION & HOW GENES WORK

GENETIC INFORMATION

The genetic information of an organism consists of a number of genes which are carried on the chromosomes found in the nucleus of every cell. Humans are thought to have about 25 000 genes, although estimates vary.

The genetic information is identical in all the body cells of an organism. This is because of the way the chromosomes are duplicated before cell division.

As an organism grows, every cell receives a copy of the genetic information found in the organism's very first cell.

This is shown in the following diagram:

Copying genetic information

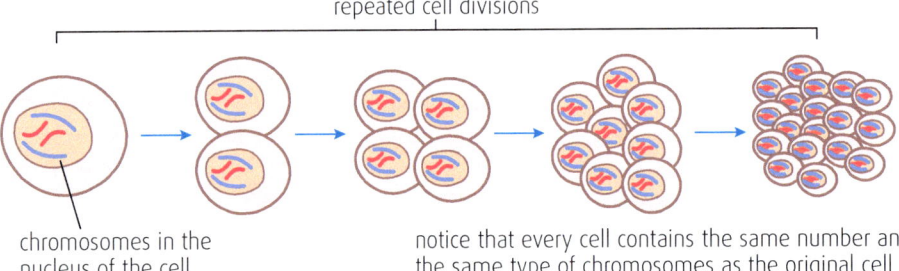

chromosomes in the nucleus of the cell

notice that every cell contains the same number and the same type of chromosomes as the original cell

Each chromosome is made from a chemical called DNA (deoxyribonucleic acid). DNA has two special properties:

- DNA forms a code for the manufacture of proteins in a cell. Proteins are important because they have many roles to play in a cell. They are part of the structure of cells and they control the chemical reactions which allow cells to develop and function. A gene is a part of a chromosome where the DNA has the code for one particular protein.
- DNA can make copies of itself (replication). This is important because it allows the chromosomes to be duplicated for cell division. In this way the genetic information can be passed from parent cell to daughter cells and also to following generations from parent to offspring.

HOW GENES WORK

DNA is a double-stranded molecule consisting of two parallel chains. Each chain carries smaller molecules called bases. There are four different bases called adenine, cytosine, guanine and thymine. They are often referred to simply as A, C, G and T.

Each of the four bases can pair up with one of the other bases in a particular way. A and T can pair together. C and G can pair together. This pairing of the bases holds the two parallel chains together. It is also the key to the two special properties of DNA – the control of protein production in a cell and chromosome replication.

Base pairing on part of a DNA molecule of a chromosome

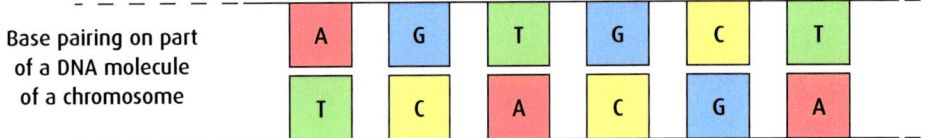

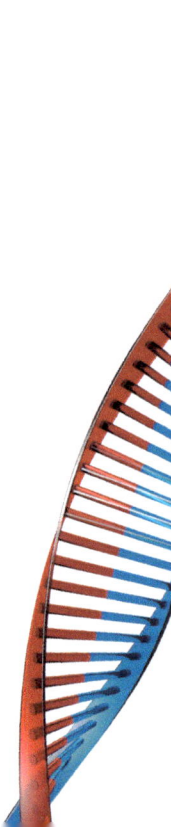

CONTROL OF PROTEIN SYNTHESIS

Protein synthesis is the way in which the genetic information of the DNA is used to control the development and the activities of a cell.

A gene is a section of a chromosome in which the sequence of bases acts as a code for the production of a protein. Different base sequences code for different proteins.

The mechanism of protein synthesis depends on the base pairing described earlier.

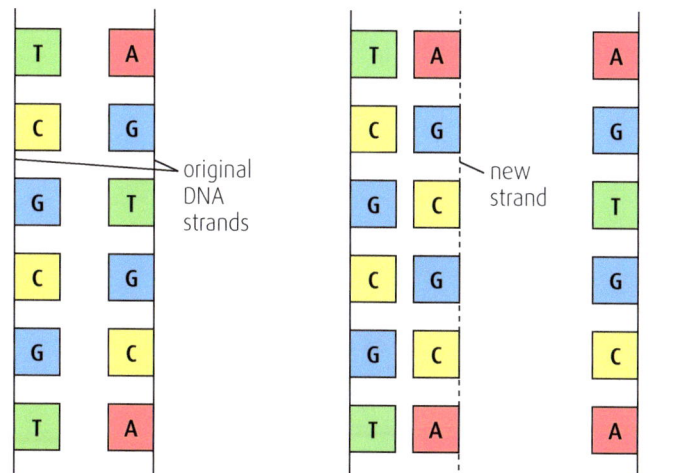
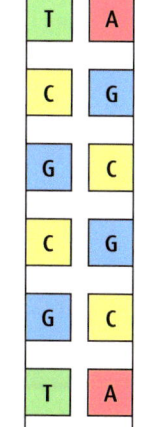

A section of the two parallel strands of the DNA molecule separate from each other

Individual free bases pair up with the exposed bases on one of the DNA strands and join up to form a single-stranded chain

The new single-stranded molecule moves from the nucleus into the cell cytoplasm. The original DNA strands re-join.

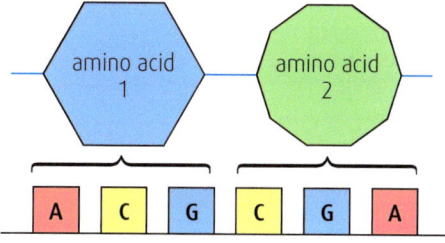

Control of protein synthesis

Each group of three bases on the new single strand acts as a code which causes individual amino acids to line up. The amino acids link up to form a chain which becomes a protein molecule.

 **DON'T FORGET**

Each gene contains the code for one particular protein. Proteins are made of amino acids. The order of DNA bases determines the order of the amino acids.

SUMMARY
- Genetic information is carried on the chromosomes in the nucleus of a cell.
- A unit of genetic information is a gene. There are many genes on one chromosome.
- A gene is a sequence of bases of the DNA on part of a chromosome.
- Each gene codes for the production of a protein in a cell.
- The sequence of bases of a gene controls the sequence of amino acids in the protein.

THINGS TO DO AND THINK ABOUT

1. The diagram shows part of the DNA of a gene.

   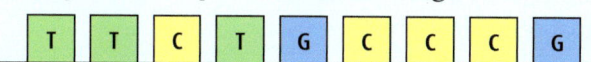

   What would be the sequence of bases on the corresponding single-stranded chain which carries this code from the nucleus when this gene is working?

2. The table gives the names of a number of amino acids together with their DNA base codes. Use the table to identify the order of amino acids which would appear in the protein produced by this part of the gene above.

DNA code	Amino acid
AAG	phenylalanine
TTC	lysine
ACG	cysteine
TGC	threonine
CCG	glycine
GGC	proline

DNA, GENES AND CHROMOSOMES

PASSING ON THE GENETIC INFORMATION

PASSING INFORMATION FROM PARENT CELL TO DAUGHTER CELLS

DNA REPLICATION

When a chromosome is duplicated, the genes which make up that chromosome are also duplicated.

The duplication of the chromosomes before cell division is due to the replication of DNA. This depends on base pairing.

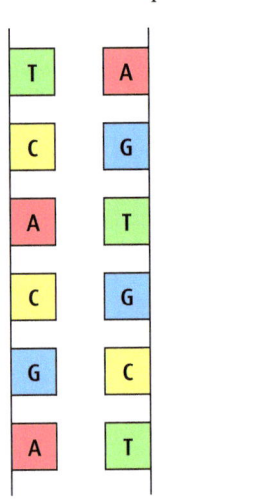
The two parallel strands of the DNA molecule separate from each other

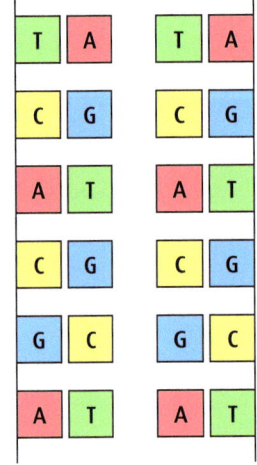
Individual free bases pair up with the bases that are now exposed on the DNA chains

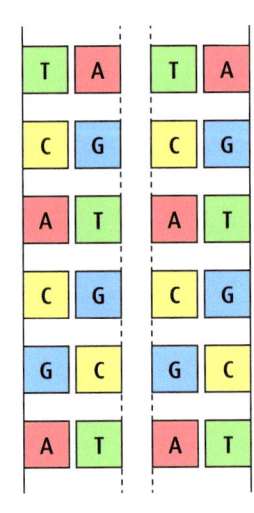
The new bases link up to form new partner strands to the original strands

DNA replication

This process results in two double-stranded DNA molecules that are identical to each other and to the original double-stranded molecule.

When this is complete along the full length of the DNA, two identical chromosomes will have been produced. Each of the daughter cells produced during cell division will receive one of the two chromosomes. This allows every cell produced to have a full set of genetic information that is identical to the information contained in the very first cell of an organism.

PASSING INFORMATION FROM PARENT TO OFFSPRING

All individuals of the same species have the same number of chromosomes in their cells. Humans have 46 chromosomes in their cells. These 46 chromosomes consist of two sets of 23 chromosomes. Each set contains one copy of every gene.

During sexual reproduction, one set of chromosomes is passed by each parent to their offspring. This happens because the sex cells contain only one set of chromosomes, rather than the normal two sets found in other body cells. This means that each parent contributes equally to the genetic information of their offspring by contributing one of every gene needed.

The following diagrams show the passing of genetic information from parents to offspring. To make the process easier to follow there are only six chromosomes shown in the body cells, rather than the 46 present in human cells. The cells of the organism involved contain two sets of three chromosomes. One set of chromosomes consists of one long, one medium and one short chromosome.

> **DON'T FORGET**
>
> The pairing of the bases is the reason why chromosomes can make copies of themselves. It is also the reason why it allows the genetic code to control the synthesis of proteins.

DNA, Genes and Chromosomes – Passing on the Genetic Information U1

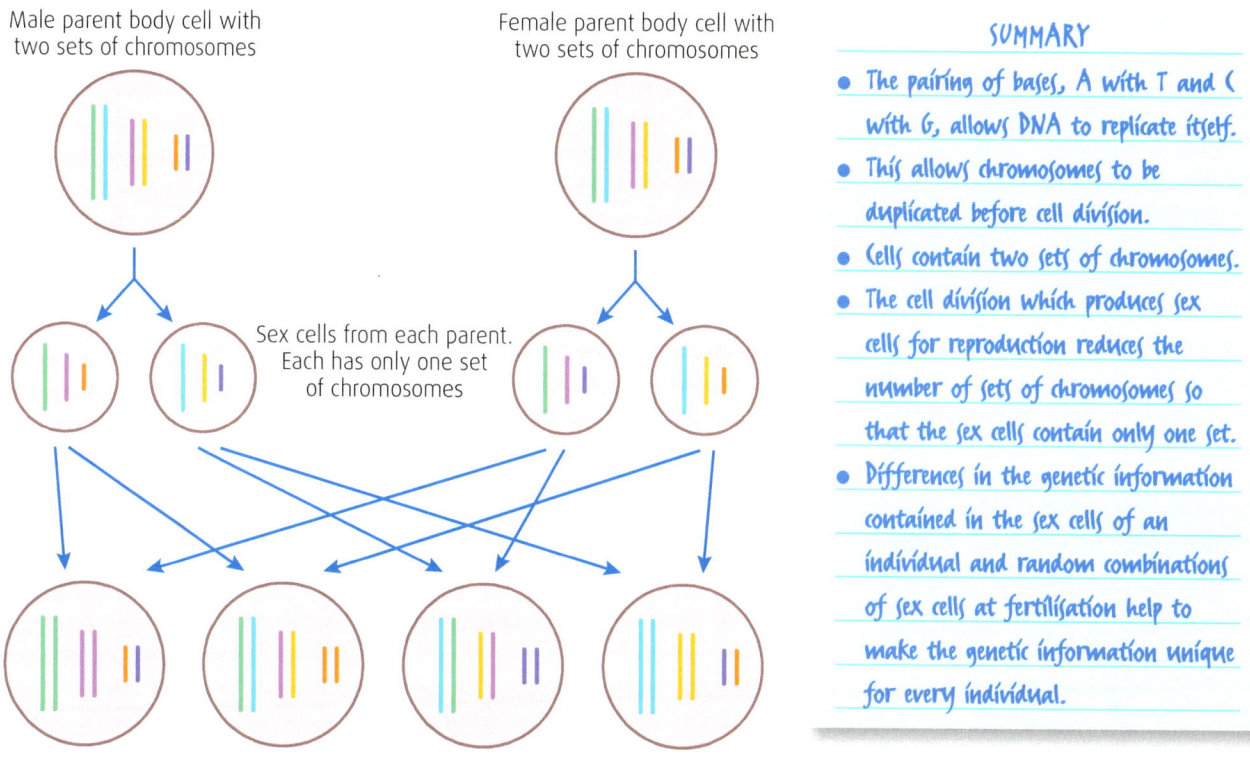

Possible results of fertilisation between one sex cell from each parent

Passing information from parent to offspring

SUMMARY
- The pairing of bases, A with T and C with G, allows DNA to replicate itself.
- This allows chromosomes to be duplicated before cell division.
- Cells contain two sets of chromosomes.
- The cell division which produces sex cells for reproduction reduces the number of sets of chromosomes so that the sex cells contain only one set.
- Differences in the genetic information contained in the sex cells of an individual and random combinations of sex cells at fertilisation help to make the genetic information unique for every individual.

A single set of chromosomes found in the sex cell could be made up of either of the two long chromosomes plus either of the two medium chromosomes and either of the two short chromosomes. This means that a parent can pass on different combinations of genetic information to their offspring. This helps to produce variation in the offspring. Variation is increased further because of the random combining of sex cells at fertilisation. This helps to make every individual unique. The only exception to this in humans is with identical twins, when the DNA is exactly the same in both twins.

THINGS TO DO AND THINK ABOUT

1. The diagram shows part of one chain of a DNA molecule.

    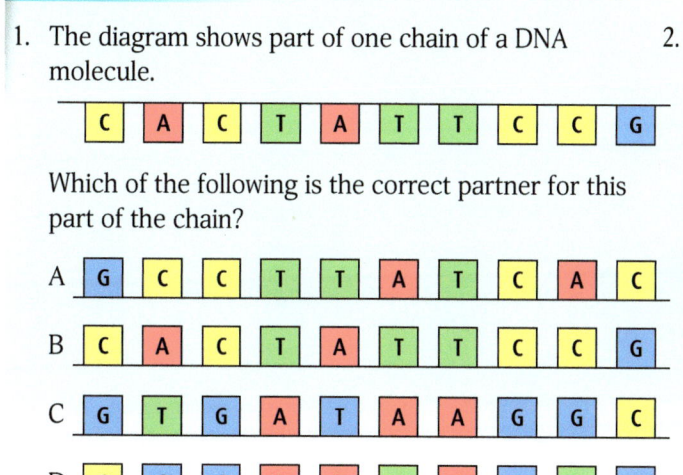

    Which of the following is the correct partner for this part of the chain?

2. The number of genetically different sex cells that can be produced increases as the number of chromosomes normally present in a single set increases. It can be calculated as 2^n, where n is the number of chromosomes in a single set.

    In the example above with three chromosomes in a single set, the number of different sex cells is 2^3 or $2 \times 2 \times 2 = 8$.

    Only four of the possible eight are shown (two from each parent).

    (a) Draw the other four possible different sex cells which could be produced by the parents.

    (b) How many different sex cells can a human produce with 23 chromosomes in a single set?

THERAPEUTIC USES OF CELLS

GENETIC ENGINEERING

Scientists have developed ways of altering the genetic information present in cells. For example, it is possible to transfer genes from one organism to another. This process is called genetic engineering.

INSULIN PRODUCTION

Insulin is a hormone that is produced by the pancreas in the body. It is essential for the control of blood sugar levels. Some people are unable to produce insulin and are said to have Type 1 diabetes. This illness is managed with a controlled diet and regular injections of insulin.

Medicinal insulin was previously obtained from slaughtered animals such as cattle and pigs. Some people show allergic reactions to animal insulin and some have ethical objections to the use of animals as a source of insulin.

It is now possible to produce human insulin from genetically modified bacteria. Genetically engineered insulin causes fewer side-effects than animal insulin. There are no ethical issues in its use and it can be produced in large quantities quickly and relatively cheaply.

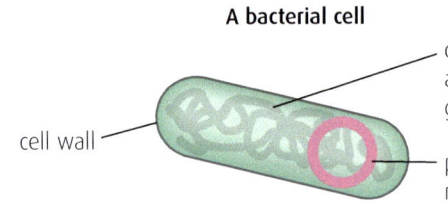

A bacterial cell
- chromosome – forms a loop and carries most of the genetic information
- cell wall
- plasmid – circular chromosomal material carrying additional genetic information

human cell showing chromosome carrying the gene for insulin production

bacterial cell

the chromosome carrying the insulin gene is removed from the cell and the insulin gene is isolated

a plasmid is removed from a bacterial cell and the loop is cut open

the insulin gene is fixed into the opened plasmid and re-joined into a loop

the altered plasmid is inserted into a bacterial cell

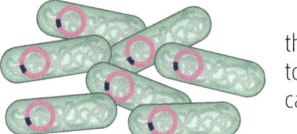

the bacterial cells reproduce to form large numbers all carrying the insulin gene

Insulin production by genetic engineering

The bacteria then produce large quantities of insulin quickly.

OTHER EXAMPLES OF THERAPETIC PRODUCTS FROM GENETIC ENGINEERING

Genetic engineering is being used to produce other important medical treatments.

EXAMPLE — Factor VIII

Factor VIII is a blood protein which is essential for the clotting action of the blood. Some people cannot produce Factor VIII and suffer from haemophilia. This means that their blood is unable to clot. They are at risk because wounds will not heal and internal bleeding continues unchecked.

Factor VIII was normally obtained from donated blood and was very limited in supply. It also had risks because of possible contamination with viruses which may have been in the blood of the donor.

Pure Factor VIII can now be produced by genetic engineering from modified bacteria.

EXAMPLE — Human Growth Hormone (HGH)

Human Growth Hormone (HGH) is responsible for the growth of children to adult height. If HGH levels are too low then a child will not grow to normal adult height.

HGH was obtained from human bodies and was very limited in supply. It can now be produced from bacteria by genetic engineering and is much more widely available.

> **SUMMARY**
> - Genetic engineering is the transfer of genes from one organism to another.
> - Human genes can be transferred into bacteria which are then able to produce human proteins. This technique is used to obtain useful products from genetically modified bacteria.

DON'T FORGET

Genetic engineering involves transferring genes from one organism to another to give the receiving organism new abilities.

THINGS TO DO AND THINK ABOUT

1. Name the structures found in bacterial cells into which the genetic material is transferred.
2. Name three human proteins that can be produced by genetic engineering to treat human disorders.
3. Give two advantages of using genetically engineered insulin rather than animal insulin in the treatment of diabetes.

THERAPEUTIC USES OF CELLS

STEM CELLS

Stem cells are unspecialised cells that can continue to divide to produce more cells like themselves as well as cells that can become specialised. The function of stem cells is to produce the cells needed for the development of different types of tissues and organs in a growing embryo. They also produce specialised cells to replace damaged or worn-out cells.

SOURCES OF STEM CELLS

Stem cells can be obtained from a number of human sources. They vary in their ability to divide continually and to produce specialised types of cells. There are ethical issues associated with the use of stem cells from some of these sources.

EMBRYONIC STEM CELLS

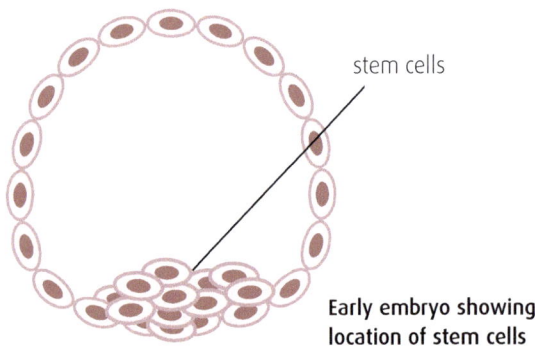

Early embryo showing location of stem cells

Embryonic stem cells are obtained from embryos which have not yet produced different types of tissues. The embryos have been created in the laboratory and have not been taken from a woman's body. Embryonic stem cells can be maintained for long periods as a cell culture and they have a very high ability to produce different types of tissues. There are serious ethical issues involved in the creation of human embryos for use in research and for medical use.

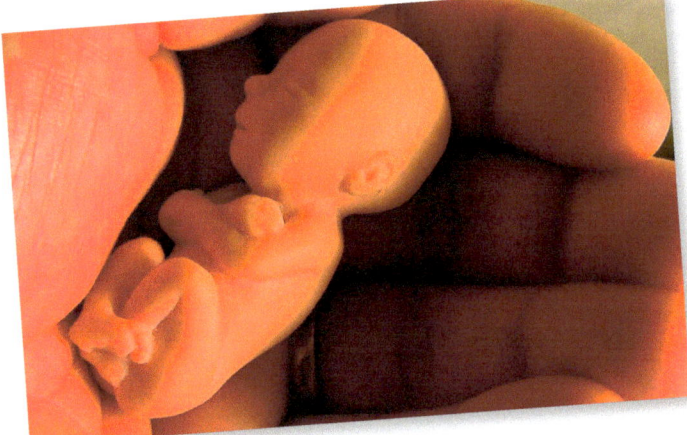

Life-size model of 12-week-old human fetus

FETAL STEM CELLS

Stem cells can also be obtained from aborted fetuses. These stem cells do not continue dividing as long as embryonic stem cells and are more limited in the range of specialised cells that they can produce. There are also ethical issues regarding the use of these cells.

UMBILICAL CORD BLOOD

Stem cells are also found in umbilical cord blood and in amniotic fluid. These stem cells are similar to fetal stem cells in their abilities. There are no ethical issues involved in their use.

Parents can store the umbilical cord blood of their children for use to treat possible future medical conditions.

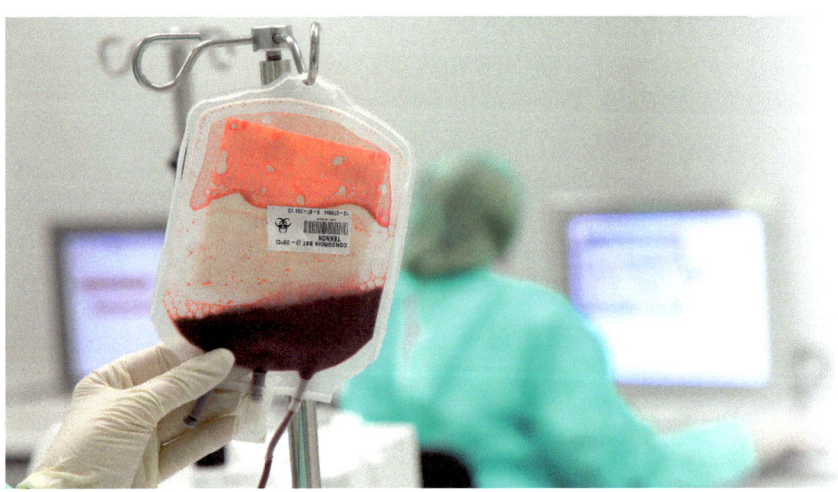

Umbilical cord blood

Therapeutic Uses of Cells – Stem Cells U1

SOME SOURCES OF TISSUE (ADULT) STEM CELLS

Stem cells can be obtained from many areas of an adult. They are similar to fetal stem cells in their abilities. Bone marrow stem cells are obtained by drilling into the thigh bone or hip bone and extracting bone marrow. Blood stem cells are obtained by drawing blood from a donor, extracting the stem cells and passing the remaining blood back to the donor. Fat stem cells are extracted from fat that is removed from the body by liposuction.

CULTURING EMBRYONIC STEM CELLS

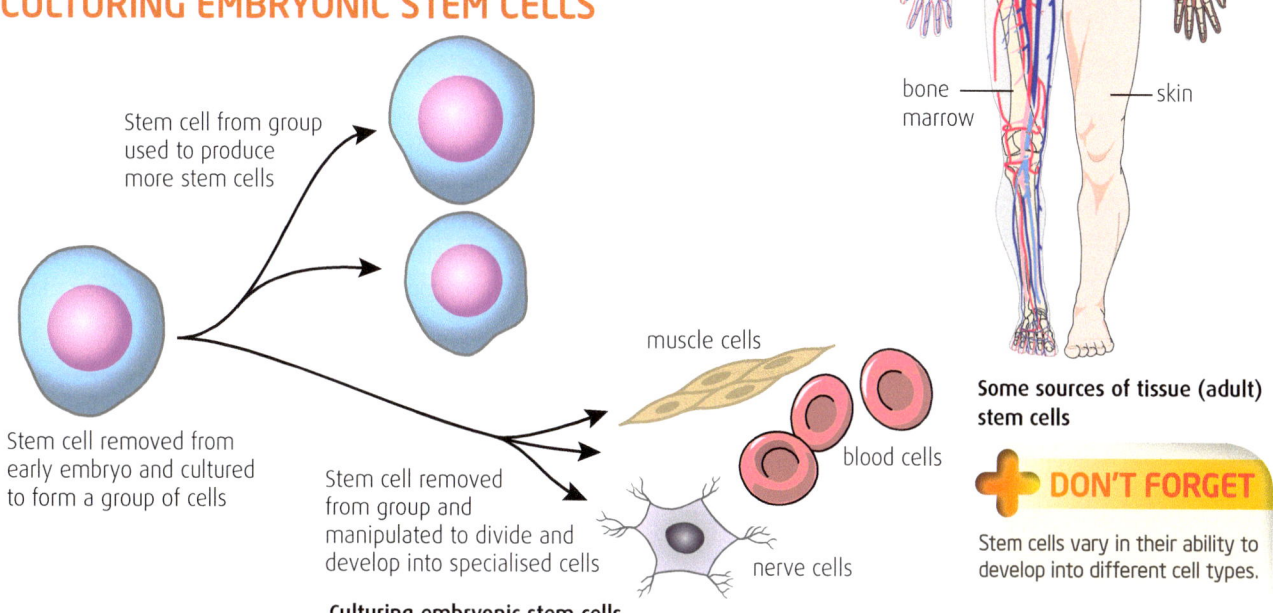

Some sources of tissue (adult) stem cells

Culturing embryonic stem cells

DON'T FORGET

Stem cells vary in their ability to develop into different cell types.

USES OF STEM CELLS

Bone marrow transplants have been used for many years to treat some forms of cancer (particularly leukaemia) and spinal cord damage.

Medical scientists are improving methods of culturing stem cells and manipulating them to increase their ability to form different cell types. It is hoped that this will enable stem cells to be used in the treatment of more and more medical conditions. These include brain damage, other cancers, heart damage, baldness, missing teeth and diabetes.

Stem cells could also be used in the testing of new drugs and the effects of possible toxins in chemicals such as pesticides.

REPLACEMENT ORGANS

Medical research has shown that stem cells can be made to develop into functioning tissues and into complete organs. The aim is to be able to produce an increasing range of tissue types and fully functioning organs for transplant into patients. Stem cells are encouraged to reproduce and grow into particular shapes to form the required tissue material or organ.

The technique has allowed the successful growth and transplant of human bladders. Other structures such as human jawbones, rat lungs and kidneys have been created in the laboratory. It is hoped that future advances in stem cell therapy will lead to the growth of functional organs for use in organ transplantation. This will help solve the problems of a shortage of donors.

SUMMARY

Stem cells are unspecialised cells. They divide to produce more stem cells and cells which can become specialised for different functions. They are essential for the development of different tissues and organs during the growth of embryos and fetuses. They also provide new cells for the repair and replacement of adult damaged or worn-out tissues.

THINGS TO DO AND THINK ABOUT

1. What is a stem cell?
2. Give two examples of sources of human stem cells.
3. Give two examples of different medical uses of stem cells.

PROPERTIES OF ENZYMES AND THEIR USE IN INDUSTRIES

PROPERTIES OF ENZYMES & ENZYMES AT WORK

PROPERTIES OF ENZYMES

BASIC FEATURES OF ENZYMES

All enzymes are made of protein. They are made in all living cells by the process of protein synthesis described earlier.

Enzymes are biological catalysts. This means that they speed up chemical reactions in living organisms and that they are unchanged at the end of the reaction. All the reactions of living organisms take place because of enzymes. Without enzymes, the essential reactions of organisms would not take place or would occur too slowly to be of use.

Enzymes are specific. This means that each enzyme can carry out only one type of reaction.

Enzyme-controlled reactions fall into two groups:

1. **Build-up reactions**
   These are reactions in which smaller molecules are joined together to build up larger molecules.

HOW ENZYMES WORK

Enzyme-controlled reactions take place on the surface of the enzyme molecule in an area called the active site. The active site of each enzyme has a particular shape which matches the shape of the molecules of the chemicals involved in the reaction. The enzyme will have no effect on chemicals with different-shaped molecules.

Chemicals involved in the reaction are called substrates and chemicals formed by the reaction are called products.

> **DON'T FORGET**
>
> SUBSTRATE $\xrightarrow{\text{ENZYME}}$ PRODUCT
>
> Human enzymes work best close to body temperature. Higher temperatures can stop them working.

2. **Breakdown reactions**
   These are reactions in which larger molecules are broken down into smaller molecules.

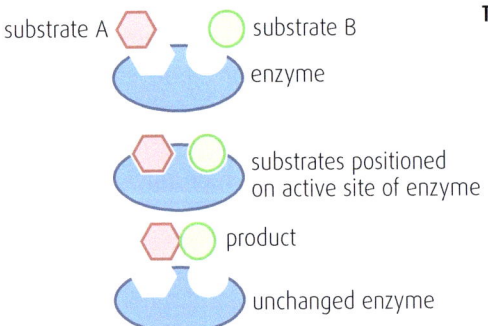
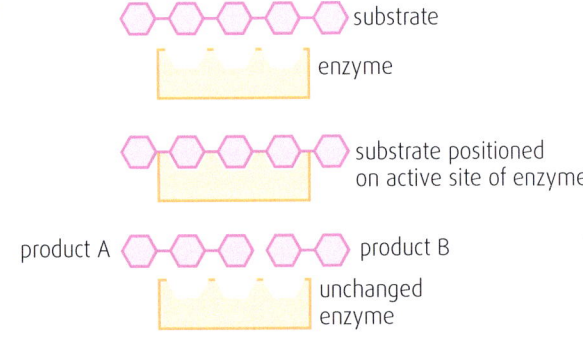

Types of enzyme reactions

ENZYMES AT WORK

TESTING SUBSTRATES AND PRODUCTS

EXAMPLE 1 — Testing for starch

Add a drop of iodine solution to a suspension of starch. A blue-black colour appears.
This only happens if starch is present.

Testing for starch

EXAMPLE 2 — Testing for sugars

Add a few drops of Benedict's reagent to a solution of sugar and heat for a few minutes. An orange-red colour appears. This only happens if a simple sugar is present.

Testing for simple sugars

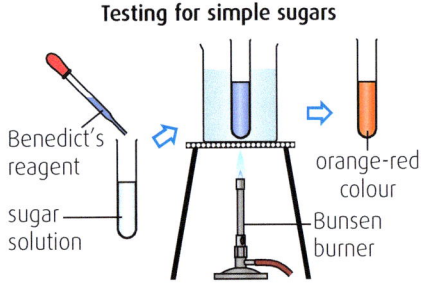

Properties of Enzymes and Their Use in Industries – Properties of Enzymes & Enzymes at Work U1

COMMON ENZYME REACTIONS

EXAMPLE 1 — Amylase enzyme

Starch is a substance found in plants where it acts as a food store. It is important in the diet of many animals, including humans. It is made of large insoluble molecules which must be broken down into smaller soluble sugar molecules which can be absorbed and used by the body.

Amylase solution can be mixed with a starch suspension and a sample of the mixture taken and tested. It would show that starch is present but not sugar.

If the mixture is left in a warm place for 30 minutes and another sample tested, the results would be different. They would show that sugar is present but that starch is not present or that there is less of it. If this was done with a starch suspension on its own, the starch would always remain and no sugar would appear. This is because the amylase enzyme is needed to break down the starch into sugar.

The action of amylase

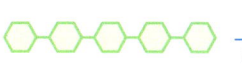 

starch (this gives a positive reaction with iodine) → amylase (a breakdown enzyme) → sugar (this gives a positive reaction with Benedict's reagent)

EXAMPLE 2 — Catalase enzyme

The cells of all organisms produce the chemical hydrogen peroxide as a waste product of some reactions. Hydrogen peroxide can cause damage to cells and so it must be removed quickly.

Catalase is important because it converts hydrogen peroxide into harmless water and oxygen. It is present in any living tissue and its activity is easily demonstrated.

A small piece of fresh animal or plant tissue, such as liver or potato, can be added to hydrogen peroxide solution. Bubbles of oxygen gas form around the tissue sample as the catalase from the cells breaks down the hydrogen peroxide.

EXAMPLE 3 — Phosphorylase enzyme

Green plants are able to make sugar which they use as a food. Sugar is not easily stored because its molecules are small and soluble.

Phosphorylase is an important enzyme because it allows plants to convert the sugar into starch which can be stored for later use. Its activity can be easily demonstrated. Phosphorylase can be obtained by mashing raw potato and filtering the mash to obtain starch-free juice. This extract contains the phosphorylase enzyme.

Glucose phosphate is a simple sugar which is used in cell reactions.

Starch-free potato extract can be mixed with a solution of glucose phosphate and a sample of the mixture taken and tested. It would show that glucose sugar is present but not starch.

If the mixture is left in a warm place for 30 minutes and another sample tested, the results would be different. They would show that starch is present but that glucose sugar is not present or that there is less of it.

If this was done with a glucose phosphate solution on its own, the glucose would always remain and no starch would appear. This is because the phosphorylase enzyme is needed to build up the glucose phosphate into starch.

SUMMARY
- Enzymes are biological catalysts which speed up the reactions of living cells.
- They are essential because the vital reactions of cells would not take place fast enough without them.

THINGS TO DO AND THINK ABOUT

1. What is meant by the following statements?
   (a) Enzymes are biological catalysts.
   (b) Enzymes are specific.
2. Which type of reaction, build-up or breakdown, is carried out by each of the following enzymes?
   (a) Amylase   (b) Phosphorylase   (c) Catalase
3. How could the procedure described to demonstrate the activity of catalase be adapted to compare the catalase content of different tissues? What precautions would you take to ensure that your results were valid?

PROPERTIES OF ENZYMES AND THEIR USE IN INDUSTRIES

THE USE OF ENZYMES IN INDUSTRY

Enzymes are used in a wide range of industrial processes. Some enzymes are extracted from organisms but there is an increasing use of enzymes that have been artificially synthesised.

EXTRACTING ENZYMES

A method for extracting enzymes from microorganisms is shown below.

Extraction of enzymes from microorganisms

Microorganisms are cultured in relatively small numbers on a dish and then in larger numbers in a flask.

They are transferred to a large-scale fermenter where they grow in large numbers and produce the enzyme.

Microorganisms are collected from the fermenter. Their cells are broken down and the enzyme is extracted and purified.

DON'T FORGET

Enzymes are useful in industry because they work quickly at relatively low temperatures. Using enzymes is efficient because they do not get used up during the reaction. This allows them to work again and again.

The photograph shows an industrial facility for extracting enzymes for industrial use.

USES OF ENZYMES

Uses for the enzymes previously described are given in the table:

Enzyme	Sources	Uses
Amylase	Fungi and bacteria	1. Breakdown of starch in flour to speed up the activity of yeast in causing the rise of bread dough. 2. Breakdown of starch in starch syrup to produce high-value sugar syrup. 3. Breakdown of starch in barley grains to speed up the activity of yeast in producing alcohol. 4. Breakdown of starchy deposits on textiles and crockery by its addition to cleaning products.
Catalase	Fungi	1. Removal of hydrogen peroxide traces in foods. The hydrogen peroxide is used to kill bacteria. 2. Production of oxygen in foods along with another enzyme as a way of removing unwanted glucose. 3. Removal of hydrogen peroxide from textiles during production. 4. Contact lens cleaner. 5. Producing foam rubber from liquid rubber (latex).
Phosphorylase	Plants	1. Research is continuing on the production of various starch-like products for biotechnological use.

Other uses of enzymes include:

- Waste-water treatment to remove toxic chemicals and organic matter.
- Genetic engineering to produce substances for the treatment of diabetes and other diseases.
- Processing treatments for fabrics and leather.
- Production of biofuels.
- Cheese making using rennet, a mixture of enzymes from the stomachs of slaughtered calves. These are being replaced with manufactured enzymes. Rennet causes the separation of milk into solid curds and liquid whey.

Curds and whey

SUMMARY

- Enzymes for industrial use are normally obtained from microorganisms.
- Some are extracted from plants or the organs of slaughtered animals.
- Some enzymes are now being produced synthetically.

THINGS TO DO AND THINK ABOUT

1. Why can enzymes carry out the same reaction again and again?
2. Explain the function of rennet in cheese making. Which of the two substances produced by the action of rennet will become the cheese?
3. Give two other uses of enzymes in industry.

PROPERTIES OF MICROORGANISMS AND THEIR USE IN INDUSTRIES

USEFUL CHARACTERISTICS OF MICROORGANISMS & TRADITIONAL USES OF YEAST

For thousands of years people have used microorganisms to make food products. Initially the people had no knowledge of the microorganisms or of the enzymes produced by them. In all cases, it is the enzymes produced by the microorganism which carry out the reactions involved.

USEFUL CHARACTERISTICS OF MICROORGANISMS

The microorganisms most commonly used are single-celled organisms, in particular various types of bacteria, and the single-celled fungus, yeast. Such microorganisms are useful for several reasons:

- they are able to carry out particular reactions which are part of the manufacturing process of some food products
- they are able to reproduce rapidly by cell division
- they can be grown on simple food sources
- they can be grown in large numbers in controlled conditions

TRADITIONAL USES OF YEAST

Yeast has the ability to survive without oxygen, although it grows better if oxygen is available. If oxygen is not available, yeast will survive by releasing energy from sugars using a process called fermentation. The products of this are carbon dioxide and alcohol. Both of these products are useful.

THE USE OF YEAST IN BREWING AND WINE MAKING

Brewing is the steeping of starchy foods such as cereal grains in water to convert the starch to sugars and then fermenting the sugary liquid with yeast to produce beer. Wine making is fermenting the sugary juice of grapes with yeast. In both cases, yeast uses the sugars as food in the absence of oxygen and produces alcohol as a result.

Today the process of fermentation is used to produce a range of alcoholic drinks.

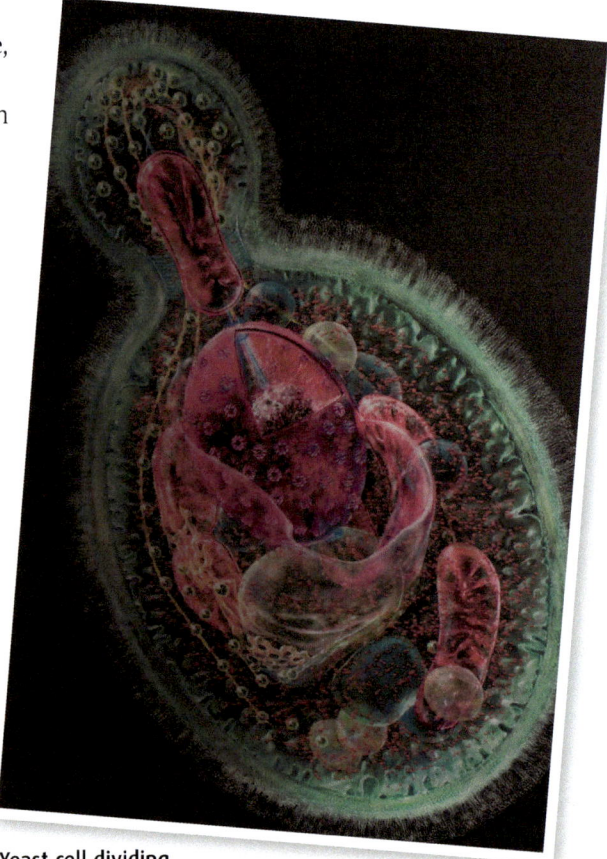

Yeast cell dividing

Properties Of Microorganisms and Their Use in Industries – Useful Characteristics of Microorganisms & Traditional Uses of Yeast | U1

Fermenting beer in a small and a large Scottish brewery

Small brewery with wooden fermenting vessels

Large brewery with stainless steel fermenting vessels

Wine making

The traditional method of crushing grapes to get the juice for winemaking

Fermenting vessels in a modern winery

The use of yeast in baking

Yeast can also be used in bread making. If sugar is included in bread dough and yeast is added, fermentation will take place. Alcohol and carbon dioxide will be produced and the dough will rise (increase in volume). It is the carbon dioxide which is important in making the dough rise because it forms bubbles which become trapped in the dough. The alcohol which is also produced is evaporated during the baking process.

DON'T FORGET

Yeast is a single-celled fungus. There needs to be a lack of oxygen for yeast to produce alcohol.

SUMMARY

- Microorganisms have been used for thousands of years in the making of some food products.
- They can survive without the need for oxygen by releasing energy from sugary food in a process called fermentation.
- When yeast ferments sugar it produces alcohol and carbon dioxide.
- The alcohol produced by yeast is used in the manufacture of alcoholic drinks such as beer and wine.
- The carbon dioxide produced by yeast makes the dough rise in the making of bread.

THINGS TO DO AND THINK ABOUT

1. Name two traditional industries which use yeast.
2. Name the process carried out by yeast in the absence of oxygen.
3. Name the two chemicals produced by yeast in the absence of oxygen.

PROPERTIES OF MICROORGANISMS AND THEIR USE IN INDUSTRIES

TRADITIONAL USES OF BACTERIA & OTHER USES OF MICROORGANISMS

TRADITIONAL USES OF BACTERIA

Some bacteria can survive without oxygen by releasing energy from foods through the process of fermentation. Unlike yeast, the normal product of bacterial fermentation is lactic acid. Although milk will go off quickly, the action of bacteria changing it into yoghurt or cheese preserves the milk so that it keeps for longer.

THE USE OF BACTERIA IN YOGHURT PRODUCTION

Natural yoghurt contains living bacteria, and a small quantity of this type of yoghurt can be used as a source of bacteria to make more yoghurt. In the commercial production of yoghurt, cultures of specific bacteria are used. These bacteria use lactose, a sugar present in milk, and produce lactic acid.

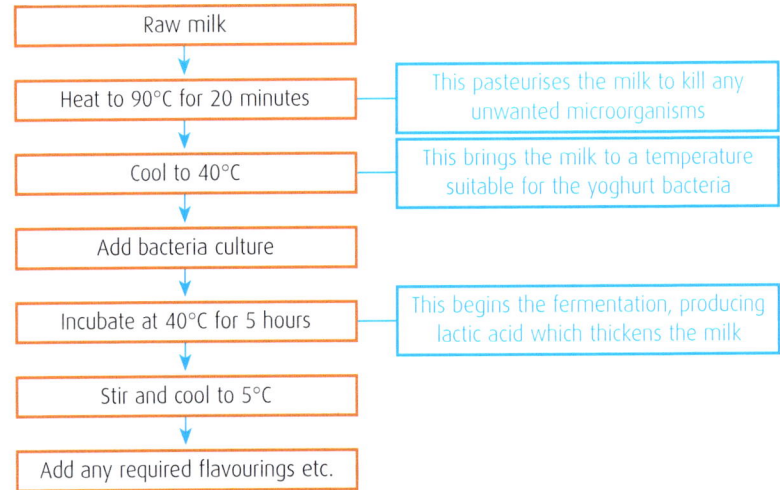

Stages in the commercial production of stirred yoghurt

THE USE OF BACTERIA IN CHEESE MAKING

Bacteria have much the same role in cheese making as they have in yoghurt production and the same types of bacteria can be used. The formation of lactic acid causes a thickening of the milk. The main difference between cheese making and yoghurt making is that rennet enzymes are used in cheese making. The rennet causes the proteins in milk to clot. This produces solid curds and liquid whey (see the photo on page 21).

Cottage cheese is curds and whey with most, but not all, of the whey drained away. For more solid cheeses the whey is completely drained away. The solid curds are then pressed into shape and left to mature.

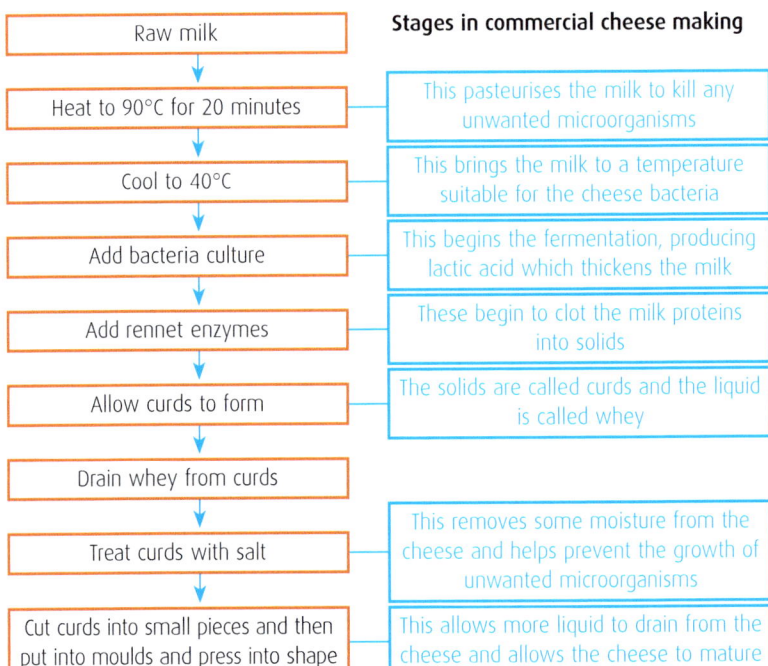

Cheese production

OTHER USES OF MICROORGANISMS

THE USE OF MICROORGANISMS IN BIOFUEL PRODUCTION

Biofuels are fuels produced by the breakdown of organic matter, usually plant material, by microorganisms such as yeast and bacteria. Yeasts produce alcohol from sugars in the same way as they do in brewing. Sugar cane and some cereals are grown for this purpose.

Bacteria are useful because they are able to break down tough plant fibres, such as plant cell walls, into sugars which the yeast can use. This allows waste plant material to be used for biofuel production.

Biofuels are a renewable alternative to fossil fuels such as petrol and diesel because more crop plants can be grown to produce more fuel. Biofuels are less harmful to the environment because the carbon dioxide they release when they are burned has recently been removed from the atmosphere by the growing plants. This means that biofuels do not add extra carbon dioxide to the atmosphere.

Biofuels are not without problems. These include:

- land being used for growing crops for biofuel rather than for food
- increased deforestation, that is, the large-scale removal of trees to create land for growing biofuel crops
- increased water usage to irrigate biofuel crops

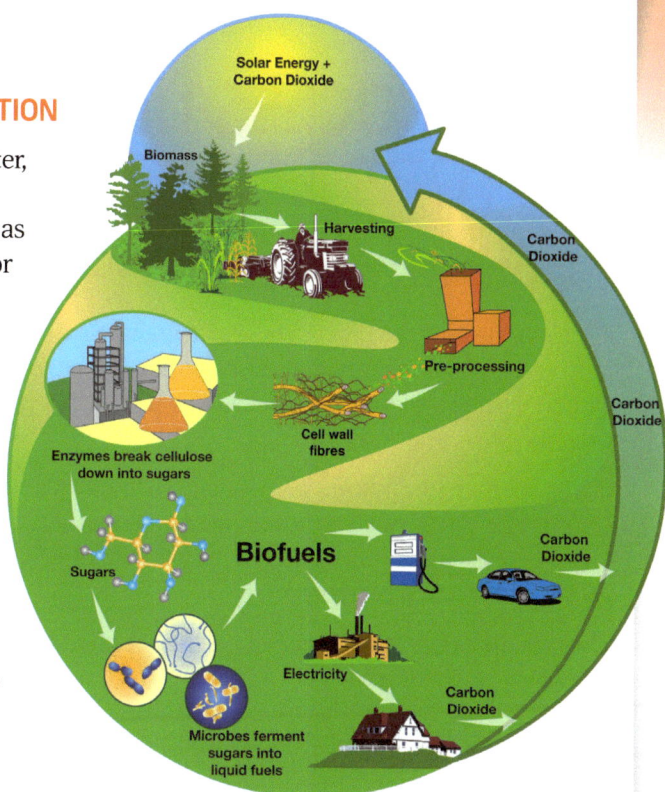

Biofuel production

OTHER USES OF BACTERIA

The importance of bacteria in the production of proteins by genetic engineering has been covered earlier (see pages 14-15).

Bacteria are also important in the breakdown of sewage to provide clean water and prevent the spread of disease.

> **SUMMARY**
> - Bacteria can release energy from sugars without the need for oxygen.
> - They produce lactic acid when they do so.
> - Bacteria are used in the production of yoghurt and cheese from milk.
> - The lactic acid they produce causes the milk to sour and thicken.
> - Microorganisms such as yeast and bacteria are used in the production of biofuels and bacteria are used in genetic engineering.

DON'T FORGET

Bacteria and yeast can both survive without oxygen but the products from each are different. Yeast produces carbon dioxide and alcohol, while bacteria produce lactic acid.

THINGS TO DO AND THINK ABOUT

1. Some bacteria can divide very quickly in ideal conditions. One bacterial cell can become two after 20 minutes. After another 20 minutes, there are four cells and after another 20 minutes there are eight cells. Starting with one cell, how many bacteria would there be after (a) 1 hour and (b) 5 hours?

2. What particular condition is needed by yeast for it to produce alcohol?

3. Cheese is made from milk. Name two other biological materials used in its production.

PHOTOSYNTHESIS

THE PROCESS OF PHOTOSYNTHESIS & THE FACTORS NEEDED FOR PHOTOSYNTHESIS

THE PROCESS OF PHOTOSYNTHESIS

Photosynthesis is the process by which green plants make their own food. They do this by using light energy to make molecules of sugar from the simpler molecules of carbon dioxide and water. Oxygen is also produced by this process and is released from the plant as a waste product.

Plants are able to carry out photosynthesis because of the green chemical chlorophyll. Chlorophyll is present in small structures called chloroplasts which are found in cells in the green parts of plants (see page 6). It is the chlorophyll which gives plants their green colour.

The process of photosynthesis can be represented by the following equation.

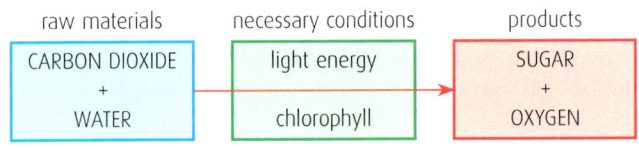

Photosynthesis

THE FACTORS NEEDED FOR PHOTOSYNTHESIS

The sugar produced by photosynthesis is the food which the plant uses for its growth and to provide the energy for cell reactions. Plants convert the sugar into starch for storage until it is needed. The presence of starch in plant leaves demonstrates that photosynthesis has taken place. By testing for the presence of starch in plants which have been kept in different conditions, the factors needed for photosynthesis can be identified.

TESTING LEAVES FOR STARCH

The presence of starch can be proven by its reaction with iodine solution (see page 18).

To carry out this test on plant leaves, a leaf must be softened and the green chlorophyll must be removed. This is to allow the iodine to enter the leaf cells and to make sure that the colour of the chlorophyll does not mask the colour change of the iodine.

The procedure for doing this is shown in the diagram.

> **DON'T FORGET**
>
> Carbon dioxide and water are the raw materials used in photosynthesis. Sugar and oxygen are the products. Light provides the energy for the chemical reactions involved and chlorophyll is needed to absorb the light so that its energy can be used.

Testing leaves for starch

Photosynthesis – The Process Of Photosynthesis & The Factors Needed For Photosynthesis — U1

THE NEED FOR LIGHT, CARBON DIOXIDE AND CHLOROPHYLL

The diagram shows a plant with variegated leaves. This means that the leaves have white areas which do not contain chlorophyll.

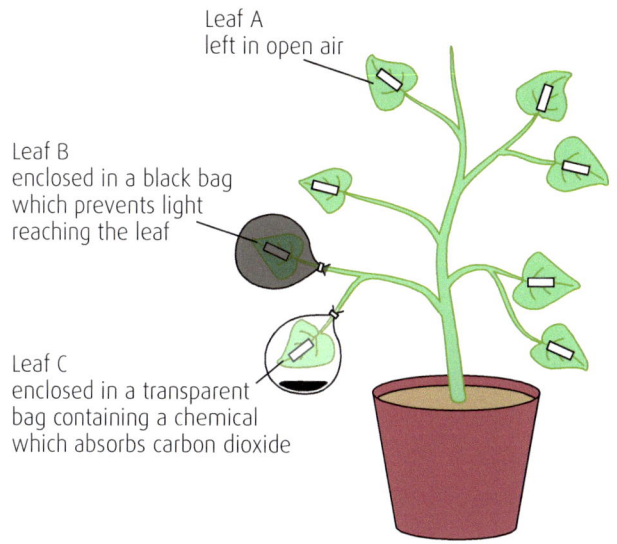

Leaf A
left in open air

Leaf B
enclosed in a black bag which prevents light reaching the leaf

Leaf C
enclosed in a transparent bag containing a chemical which absorbs carbon dioxide

Leaf	Conditions available to leaf	Result of starch test ■ positive ■ negative
A	light carbon dioxide chlorophyll in green area	positive
B	carbon dioxide chlorophyll in green area	negative
C	light chlorophyll in green area	negative

The results of testing leaves A, B and C for the presence of starch show that light, carbon dioxide and chlorophyll are all essential for photosynthesis. If any one of these factors is missing, photosynthesis will not take place.

SUMMARY

- Green plants make their own food by photosynthesis. They absorb light using chlorophyll and use the energy to combine carbon dioxide with water to produce sugar and oxygen.
- The oxygen is a waste product which is released into the air. The sugar can be stored as starch until it is needed by the plant.
- Starch can be detected in plant leaves using iodine solution. The leaves must be treated to allow the iodine to enter the leaf cells and to remove the green chlorophyll before the iodine is used.

THINGS TO DO AND THINK ABOUT

1. Name the raw materials of photosynthesis.
2. Name the source of energy used by plants for photosynthesis and say how this energy is made available to plants.
3. Describe how the leaves of green plants are treated so that they can be tested for the presence of starch.

PHOTOSYNTHESIS
MEASURING THE RATE & LIMITING FACTORS

MEASURING THE RATE OF PHOTOSYNTHESIS

The rate of photosynthesis can be obtained by measuring the average number of bubbles of oxygen released from an aquatic plant in a known period of time. The method is shown in the diagram.

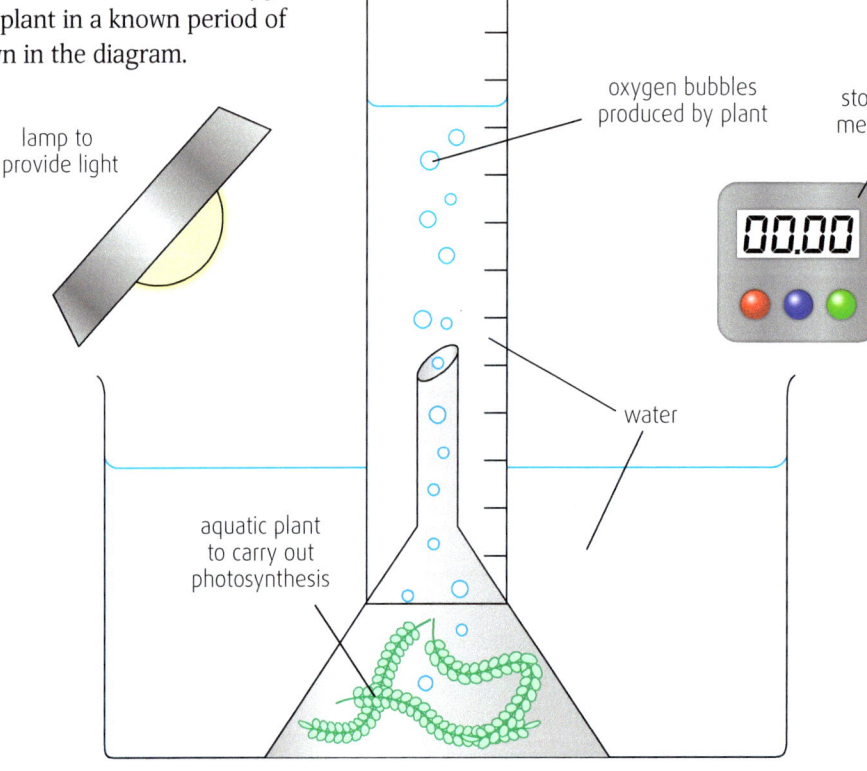

Measuring the rate of photosynthesis

If any one of the conditions required for photosynthesis is in short supply, the rate at which photosynthesis takes place will be reduced.

THE EFFECT OF LIGHT INTENSITY ON THE RATE OF PHOTOSYNTHESIS

To investigate the effect of light intensity on the rate of photosynthesis the average number of bubbles produced in a given time would be measured for a range of different light intensities. This can be done using a lamp with a dimmer control, or by altering the distance between the lamp and plant.

The results are then plotted on a graph. All other factors must be kept constant during the procedure. The results of such an investigation are shown on the graph.

This shows that at very low light intensities there is no photosynthesis. As the light intensity increases so does the rate of photosynthesis. Over this range of light intensities (A to B), light is the limiting factor for photosynthesis because it is the light intensity that is controlling the rate. If more light is given, the rate increases. However, as the light intensity increases it reaches a point where further increases (B to C) have no effect on the rate. This is because one of the other essential factors for photosynthesis is now limiting the rate and no increase in the rate is possible until the supply of that factor is increased.

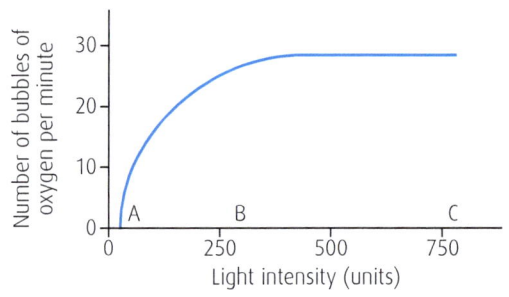

Effect of light intensity on the rate of photosynthesis.

Photosynthesis – Measuring the Rate & Limiting Factors U1

THE EFFECT OF CARBON DIOXIDE CONCENTRATION ON THE RATE OF PHOTOSYNTHESIS

The effect of carbon dioxide concentration on the rate of photosynthesis can be investigated in a similar way. In this case, the carbon dioxide concentration can be varied by placing the aquatic plant in different concentrations of sodium hydrogen carbonate solution. This provides dissolved carbon dioxide to the plant. All other factors must be kept constant.

The graph shows the results of such an investigation.

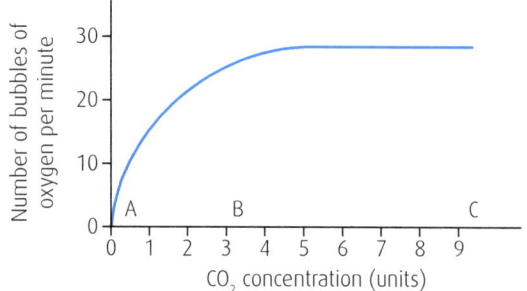

Effect of carbon dioxide concentration on the rate of photosynthesis

At low concentrations carbon dioxide can be seen to be the limiting factor for the rate of photosynthesis. Over this range of concentrations (A to B), carbon dioxide is the limiting factor for photosynthesis because it is the carbon dioxide concentration that is controlling the rate. If more carbon dioxide is available, the rate increases. As the carbon dioxide concentration increases a point is reached where further increases (B to C) make no difference to the rate of photosynthesis. At this point, carbon dioxide availability stops being the limiting factor and one of the other factors is now the limiting factor.

THE EFFECT OF TEMPERATURE ON THE RATE OF PHOTOSYNTHESIS

The effect of temperature can be investigated by carrying out the procedure using a water bath to control the temperature of the water containing the aquatic plant. Again, all other factors must be kept constant.

The graph shows the results.

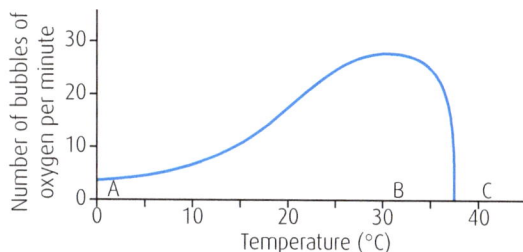

The effect of temperature on the rate of photosynthesis

The pattern of the graph is different from those of the other factors. At low temperatures, an increase in temperature causes an increase in the photosynthesis rate. Over this range of temperatures (A to B), temperature is the limiting factor for photosynthesis because it is the temperature that is controlling the rate. A point is reached where further increases in temperature do not produce increased photosynthesis. Instead, the rate of photosynthesis decreases and falls to zero (B to C). This is because the higher temperatures have a damaging effect on the structure of the protein enzymes which carry out the reactions of photosynthesis. This damage is permanent and reducing the temperature will not reverse the damage.

LIMITING FACTORS

The investigations described above show that the essential factors of photosynthesis can each limit the rate of photosynthesis if in short supply. If the supply of that factor increases, it will cease to be the one on shortest supply and so it will no longer be the limiting factor.

 **DON'T FORGET**

A limiting factor is a requirement for a process which is in short supply. It is this factor which determines the speed of the process.

SUMMARY

A limiting factor is something which restricts the rate of a process because it is in short supply. If a process has a number of factors which are essential, only one of them will be limiting the rate of the process at any time. If the supply of that factor is increased, the rate of the process will increase. However, a point will be reached when an increase in the supply of that factor will not result in an increase in the rate of the process. This is because one of the other factors has become the limiting factor.

THINGS TO DO AND THINK ABOUT

1. Why is it easier to measure the rate of photosynthesis using an aquatic plant rather than a land plant?
2. Name three factors which can be investigated for their effect on the rate of photosynthesis.
3. Water is essential for photosynthesis. Explain why the method described above is not suitable for investigating the effect of water on photosynthesis.

RESPIRATION

THE PROCESS OF RESPIRATION & RESPIRATION WITH AND WITHOUT OXYGEN

THE PROCESS OF RESPIRATION

Respiration occurs in all living organisms. It is the process by which living cells release energy from food. The food material normally involved is the sugar, glucose. The energy released by respiration is used to carry out the chemical reactions that are essential for the growth and functioning of cells.

The most efficient form of respiration involves the use of oxygen, although it is possible for cells to carry out respiration in the absence of oxygen.

RESPIRATION WITH OXYGEN

If oxygen is available, cells are able to break down glucose into carbon dioxide and water. This is the most efficient form of respiration, allowing the maximum amount of energy to be released from the glucose.

Land living animals and plants obtain oxygen from the air around them. Aquatic organisms such as fish and seaweeds use dissolved oxygen from the surrounding water.

The process of respiration can be represented by the following equation.

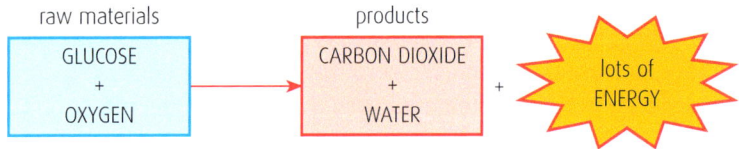

RESPIRATION WITHOUT OXYGEN

If oxygen is unavailable, organisms are still able to release energy from glucose. This process is less efficient than respiration with oxygen because less energy is released from the glucose. Also, it may only be possible for a limited time in some organisms. Some other organisms however can survive indefinitely in the absence of oxygen.

There are two different types of respiration without oxygen. These take place in different types of organisms.

ANIMALS

If the demand for energy is high, animal muscle cells can release energy from glucose by breaking it down into lactic acid. This process does not require oxygen but it releases less energy than if oxygen were used. This can only happen for a limited time because the lactic acid causes the muscle to ache and stop working. This is called muscle fatigue.

Respiration in animal cells without using oxygen is represented by the following equation.

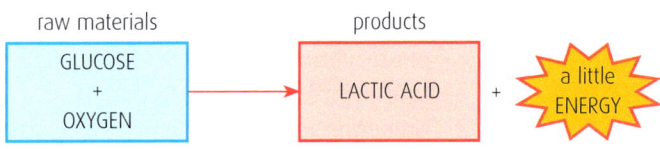

This process is similar to that which takes place in bacteria during the production of yoghurt and cheese (see page 24).

DON'T FORGET

Respiration using oxygen is more efficient than respiration without oxygen. More energy is released from each molecule of glucose used. The word equation for respiration with oxygen is the opposite of the word equation for photosynthesis.

Respiration – The Process of Respiration & Respiration With and Without Oxygen — U1

PLANTS AND YEAST

In the absence of oxygen, plant cells and fungi such as yeast are able to release energy from glucose by breaking it down into alcohol and carbon dioxide. This process is less efficient than respiration using oxygen.

Respiration in plant and yeast cells without using oxygen is represented by the equation (right).

This process is similar to that which takes place in yeast during brewing, wine making and baking (see pages 22–23).

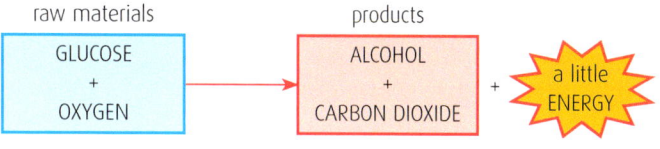

INVESTIGATING FEATURES OF RESPIRATION

PROVING THAT CARBON DIOXIDE IS PRODUCED BY LIVING ORGANISMS

Bicarbonate indicator changes colour if the concentration of carbon dioxide increases or decreases. If the concentration increases, the colour change is from red to yellow. If the concentration decreases, the colour change is from red to purple.

If the apparatus shown is set up and left, the bicarbonate indicator changes from red to yellow, showing that living plant and animal tissues produce carbon dioxide.

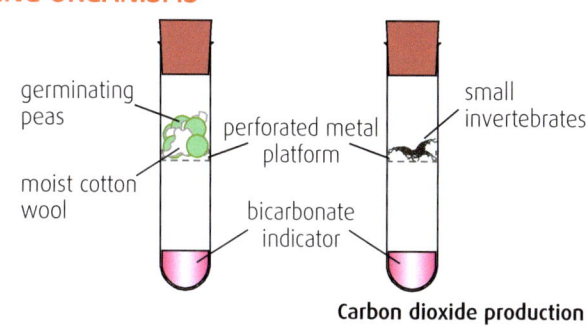

Carbon dioxide production

PROVING THAT OXYGEN IS USED BY LIVING ORGANISMS

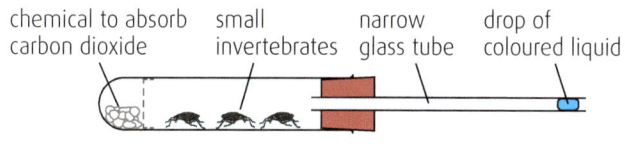

Measuring oxygen uptake

The organisms in the test tube use up oxygen and produce carbon dioxide. The volumes of the two gases tend to balance each other. The chemical in the test tube absorbs any carbon dioxide produced and this means that the volume of air in the test tube will decrease. The decrease in volume will cause the drop of coloured liquid to move along the glass tube towards the test tube. If the narrow glass tube has a scale then the actual volume of oxygen used can be measured.

By using germinating peas instead of small invertebrates, respiration in plants can be demonstrated.

> **SUMMARY**
> - Respiration takes place in all living cells. It is the process by which the cells of all organisms obtain the energy they need to function.
> - The most efficient form of respiration uses oxygen to break down molecules of glucose into carbon dioxide and water. Energy is released from the glucose when this happens.
> - Energy can be released from glucose molecules in the absence of oxygen, but less energy is released. Animal cells and bacteria produce lactic acid when this happens. Plants and yeast produce alcohol and carbon dioxide.

THINGS TO DO AND THINK ABOUT

1. In the investigation to show that carbon dioxide is produced by living organisms, explain
   (a) what a suitable control experiment would be.
   (b) why it would be a sensible precaution to wrap the tube containing germinating peas with a light-proof cover.
2. The investigation to show that oxygen is used by living organisms was carried out using larger apparatus containing a mouse. Give a reason why using a warm-blooded animal like a mouse caused the drop of coloured liquid to move in the opposite direction from the one expected.

RESPIRATION

FACTORS AFFECTING THE RATE OF RESPIRATION

The apparatus used to show that oxygen is absorbed during respiration can be adapted to measure the rate of respiration.

INVESTIGATING THE EFFECT OF TEMPERATURE ON THE RATE OF RESPIRATION

Here the tube containing the respiring organisms is upright so that it can be placed in a water bath where the temperature can be controlled. The coloured liquid is drawn up the narrow glass tube from a beaker.

The apparatus is left to stand for some time at the required temperature before any readings are taken. This allows the air and peas inside the tube to reach the temperature of the surroundings. The time for the coloured liquid to rise a particular distance (for example 5 mm) up the narrow glass tube is then recorded. Several measurements should be taken and an average calculated.

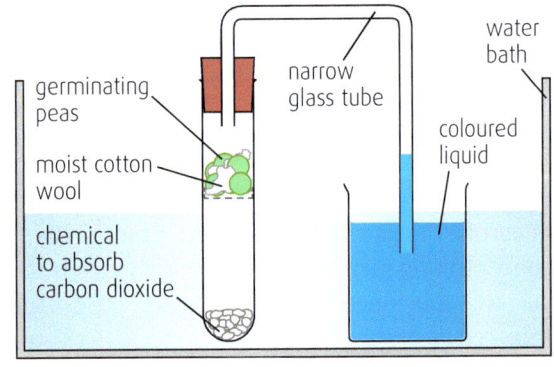

Measuring the effect of temperature on the rate of respiration

This procedure should be followed for a number of different temperatures over a range.

If the results are plotted, a graph similar to the one below should be obtained.

At low temperatures, an increase in temperature causes an increase in the respiration rate. Temperature is the limiting factor. A point is reached where further increases in temperature do not produce increased respiration. Instead, the rate of respiration decreases and falls to zero. This is because the higher temperatures have a damaging effect on the structure of the protein enzymes which carry out the reactions of respiration. This damage is permanent and reducing the temperature will not reverse the damage.

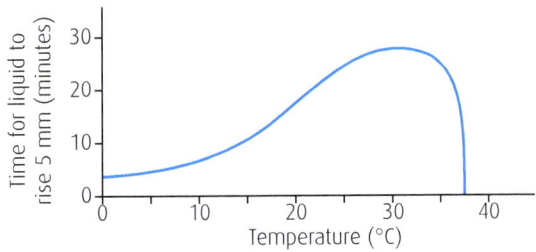

Effect of temperature on the rate of respiration

INVESTIGATING THE EFFECT OF GLUCOSE CONCENTRATION OF THE RATE OF RESPIRATION IN YEAST

In this investigation a different method of measuring the rate of respiration is used.

Instead of using movement of a coloured liquid as a measure of oxygen uptake, the production of carbon dioxide is measured by counting bubbles.

The apparatus is left to stand for some time at the required temperature before any readings are taken. This allows the yeast and glucose solution inside the tube to reach the required temperature. The number of carbon dioxide bubbles produced in a set time (for example 5 minutes) is then recorded. Several measurements should be taken and an average calculated.

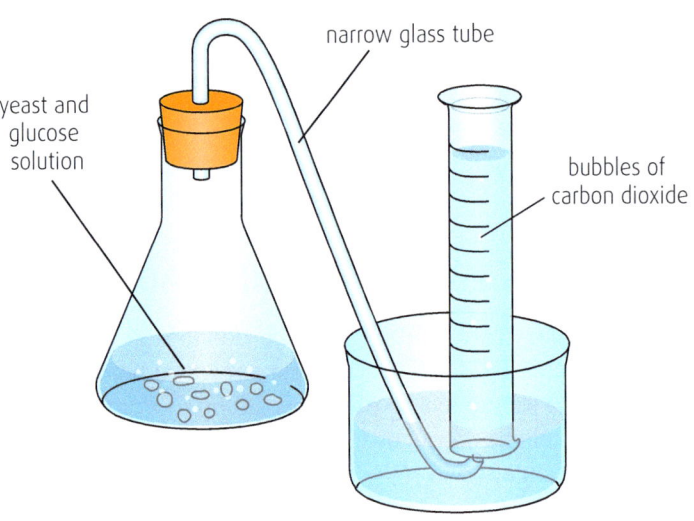

Measuring the effect of glucose concentration on the rate of respiration

This procedure should be followed for a number of different glucose concentrations over a range. The mass of yeast in the solutions and the temperature should be kept the same each time.

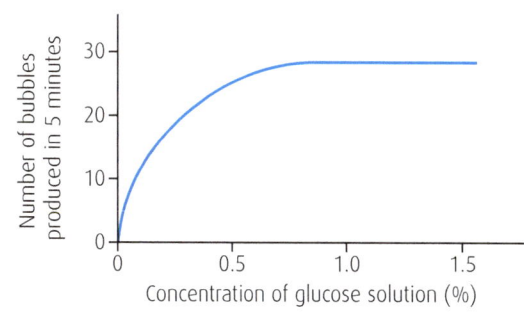

If the results are plotted, a graph similar to the one below should be obtained.

At low concentrations, glucose can be seen to be the limiting factor for the rate of respiration. If more glucose is available, the rate increases. As the glucose concentration increases, a point is reached where further increases make no difference to the rate of respiration. At this point, glucose availability stops being the limiting factor and a different factor is now limiting the rate of respiration.

SUMMARY

- In these investigations the rate of respiration is measured either as the rate at which carbon dioxide is produced by the respiring organisms, or as the rate they use oxygen.
- To measure the rate of oxygen use, the carbon dioxide they produce must be absorbed by a suitable chemical.

DON'T FORGET

When carrying out investigations on the effect of a particular factor, all other factors must be kept the same for a valid conclusion to be drawn. The results of an investigation are made more reliable when several results are taken and an average value calculated.

THINGS TO DO AND THINK ABOUT

The graph shows the results of an investigation into the effect of temperature on the rate of respiration of yeast.

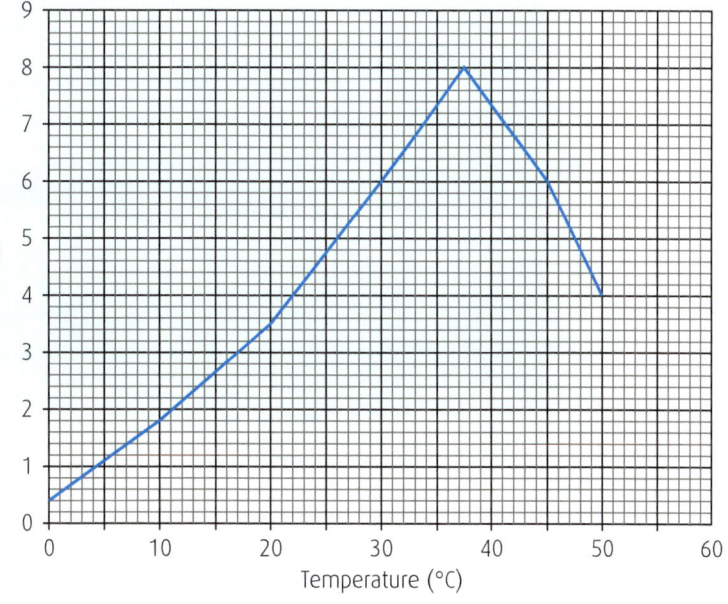

Use the information in the graph to answer the following questions:

1. At what temperature was the rate of respiration the highest?
2. Over which of the following temperature increases did the rate of respiration increase the most?

   5°C – 10°C   15°C – 20°C   25°C – 30°C   40°C – 45°C

3. Describe the effect of increasing temperature on the rate of respiration.
4. (a) As the temperature increases, at what temperature would respiration stop?
   (b) Give a reason why respiration slows and stops as the temperature increases.

CONTROVERSIAL BIOLOGICAL PROCEDURES

GENE THERAPY, PHARMING AND TRANSGENIC ORGANISMS

Biology has always involved ethical debates in which the rights and wrongs of scientific research are argued. Modern advances in knowledge about the workings of cells and the ability to manipulate the genetic information contained in cells have increased the arguments.

Genetic engineering is now well established as a technique in the production of human insulin and other medicines by modified microorganisms.

Continued scientific research means that genetic engineering is now leading towards more complex procedures including the creation of transgenic plants and animals, pharming, cloning, gene therapy and genetically modified children. Each of these procedures carry associated ethical issues which must be considered when decisions are made on whether the procedures should be allowed or not.

GENE THERAPY

Gene therapy is an experimental technique that uses genes to treat or prevent disease. In the future it may be possible to treat a disorder by inserting a gene into a patient's cells instead of using drugs or surgery.

One method of gene therapy is shown in the diagram.

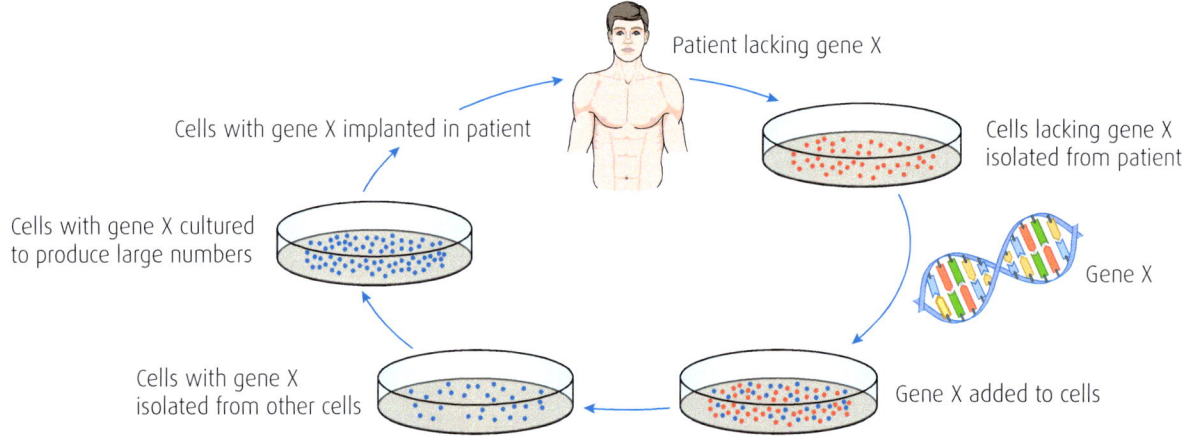

If the technique becomes widespread, it could be used for cosmetic treatments. The possibility raises many ethical questions:

- Should gene therapy be used to 'improve' basic characteristics such as height or intelligence?
- Will the high costs of gene therapy make it available only to the wealthy?
- Could the widespread use of gene therapy make society less accepting of people who are different?

Gene therapy has been used to alter particular cells of individuals by replacing defective genes with functional versions. This could lead to 'designer babies' by altering the genetic information in an embryo. If the genetic information in an egg or sperm cell is altered, it would mean that the altered information could be passed on to future individuals who had no say in whether to receive this treatment. The procedure may also affect the development of an embryo in unforeseen ways.

Controversial Biological Procedures – Gene Therapy, Pharming and Transgenic Organisms U1

PHARMING AND TRANSGENIC ORGANISMS

Pharming is a word made up from 'farming' and 'pharmaceutical'. It means the use of genetic engineering to insert new genes into animals or plants. These genes code for the production of useful chemicals such as medicines. The genetically modified or transgenic organism can then produce the chemical.

A transgenic mammal could produce the medicine in its blood, urine or milk which would be collected. The medicine would then be extracted. A method is shown in the diagram.

DON'T FORGET

Altering the genetic information in a sex cell has long-term implications. This is because if that change becomes part of a new individual, it can be passed on to following generations. Changes to the genetic information of an individual are not passed on.

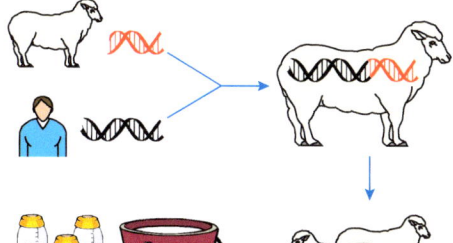

Human gene for desired protein isolated and joined to isolated sheep chromosome

Altered chromosome inserted into sheep egg which is then implanted into a sheep

Transgenic sheep produce the human protein in their milk

Offspring born with altered chromosome which can be inherited by their offspring

Pharming animals

Genetically engineered plants could produce chemicals. The plants would be harvested and the chemicals extracted from their tissues. A method is shown in the diagram.

The use of plants and animals could allow the production of chemicals that are too complex to be artificially synthesised or to be produced by genetically modified bacteria. It would be relatively inexpensive once the transgenic organisms were created.

The process has some associated ethical issues:

- Although not intended as food, transgenic animals and plants could be a risk if they entered the human food chain.
- Modified plants could cross-pollinate with unaltered plants and enter the human food chain.
- Modified plants could pose a threat to environmental biodiversity.

It is estimated that pharming is expected to be worth $100 billion globally by 2020.

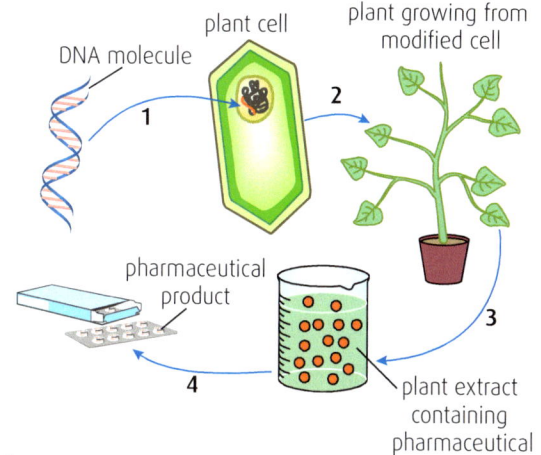

Pharming plants

SUMMARY

- The ability to alter genetic information can be used in many beneficial ways but it raises many ethical questions, especially so when it involves manipulating the genetic information of humans.
- Gene therapy may offer a method of treating inherited conditions by the replacement of faulty genes.
- Transgenic organisms have genes from different organisms inserted into them. This gives them abilities and characteristics which they would not otherwise have. These characteristics can be inherited by their offspring.
- Pharming refers to the creation of transgenic plants or animals so that they can produce useful pharmaceutical chemicals.

THINGS TO DO AND THINK ABOUT

1. Explain why the alteration of the genetic information in a human sex cell may be considered to be more controversial than altering the information in a cell of an individual?
2. Human insulin has been obtained from genetically modified bacteria for many years.
   Why is it considered necessary to use genetically modified animals and plants for the production of other chemicals?
3. Explain why the term 'pharming' is a suitable combination of the words 'pharmaceutical' and 'farming'.
4. Give one possible advantage and one possible disadvantage of the process of pharming.

CONTROVERSIAL BIOLOGICAL PROCEDURES

CLONING AND ANIMAL RESEARCH

CLONING

Cloning is the production of genetically identical copies of an organism or a cell. An animal such as a sheep is cloned by removing the nucleus from a cell of one sheep. An egg cell from a different sheep then has its nucleus removed and replaced with the nucleus from the first sheep. The altered egg cell is implanted into the second sheep. When the sheep gives birth, the lamb will be genetically identical to the first sheep. This is illustrated in the diagram.

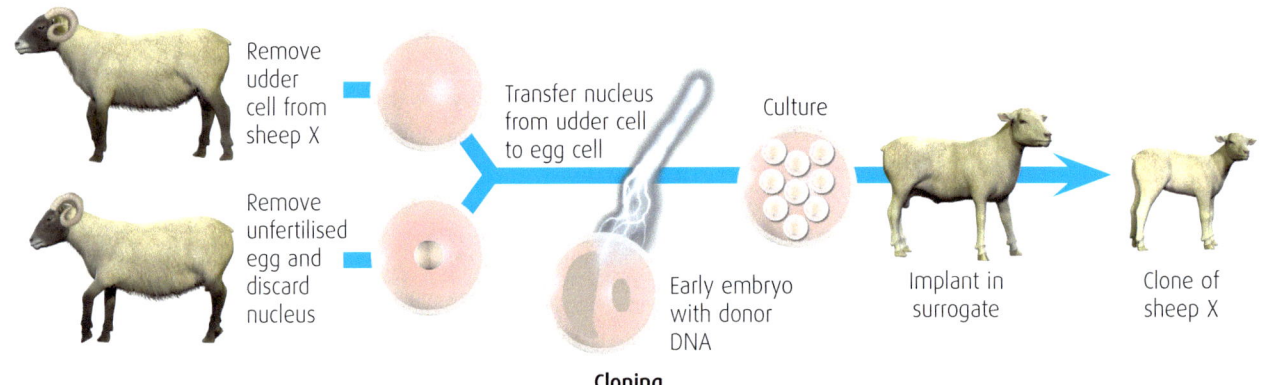

Cloning

The very first mammal to be cloned from a normal body cell was a sheep called Dolly. The same Scottish researchers who cloned Dolly have cloned other sheep that have been genetically modified to produce milk that contains a human protein essential for blood clotting. It is hoped that this will lead to the production of the protein as a medical treatment for humans.

It is also hoped that cloning will be able to produce animals with features such as good meat quality and high milk yields for agricultural purposes.

Cloning could also be used to increase populations of endangered species.

Cloning of embryos allows embryonic stem cells to be obtained for research into treatments for disease. The cloning of human embryos for this purpose is illegal.

Points to be considered about cloning include:

- Cloning whole animals is inefficient. Most cloned embryos fail to develop and, of those that do, many show defects and abnormalities.
- Cloning produces individuals that are genetically identical. Cloned populations of endangered species would lack the variability that is important for the survival of a species.
- Cloning embryos to obtain stem cells leads to the destruction of the embryos. Many people feel that this is wrong.

The issues involved with cloning would become even more serious if the techniques are ever used with human embryos.

DON'T FORGET

Cloning reduces the variability within a species. This can have disadvantages.

36

Controversial Biological Procedures – Cloning and Animal Research U1

ANIMAL RESEARCH

Animals have been used for many years in medical research. They are used to test the safety of medical products and in the development of new medicines and medical procedures. Most medicines used with humans have been developed by testing on animals. This is also true for surgical techniques and diagnostic procedures. Animals were also used for non-medical purposes such as testing the effects of cigarette smoke and cosmetics. Such testing is now banned in the European Union. It is accepted that experimenting on animals causes them harm.

People who are in favour of the use of animals in research argue that it is acceptable if the animals' suffering is kept to a minimum and that the benefits gained could not be obtained any other way.

People who are not in favour argue that no animal suffering can be justified and that there is always another way.

Choose Cruelty Free
choosecrueltyfree.org.au

Campaign posters for and against animal research

SUMMARY

- Cloning is the production of cells or organisms that are genetically identical to a parent cell or organism. Cloning mammals involves replacing the nucleus of an egg cell with the nucleus of a cell from a donor animal. The altered egg cell is then implanted into a female animal to develop. It results in an offspring that is genetically identical to the donor animal.
- The use of animals for the testing of medicinal products and procedures has provided beneficial results. It is acknowledged that such testing causes harm to the animals.
- Regulations concerning the use of animals are stricter now than in the past.

THINGS TO DO AND THINK ABOUT

1. Explain why a cloned sheep such as Dolly contains no genetic information from the sheep which gave birth to it.
2. Give one medical and one non-medical potential use of cloning mammals.
3. People in favour of the use of animals in testing say it should be allowed if two conditions apply. What are these conditions?
4. Give two arguments used by people who are against the use of animals in testing.

UNIT 2: MULTICELLULAR ORGANISMS

SEXUAL AND ASEXUAL REPRODUCTION

DIFFERENCES BETWEEN AND THE ADVANTAGES OF SEXUAL AND ASEXUAL REPRODUCTION

DIFFERENCES BETWEEN SEXUAL AND ASEXUAL REPRODUCTION

The word 'asexual' means 'without sex'. Asexual reproduction therefore refers to methods of reproduction that do not involve the essential features of sexual reproduction.

SEXUAL REPRODUCTION

Sexual reproduction is used by all types of living organisms from the simplest to the most complex.

The two most important features of sexual reproduction are:

- the production of sex cells containing only one set of chromosomes
- the joining of two sex cells at fertilisation to form one cell with two sets of chromosomes.

There are many different methods of sexual reproduction, but they all show these two essential features.

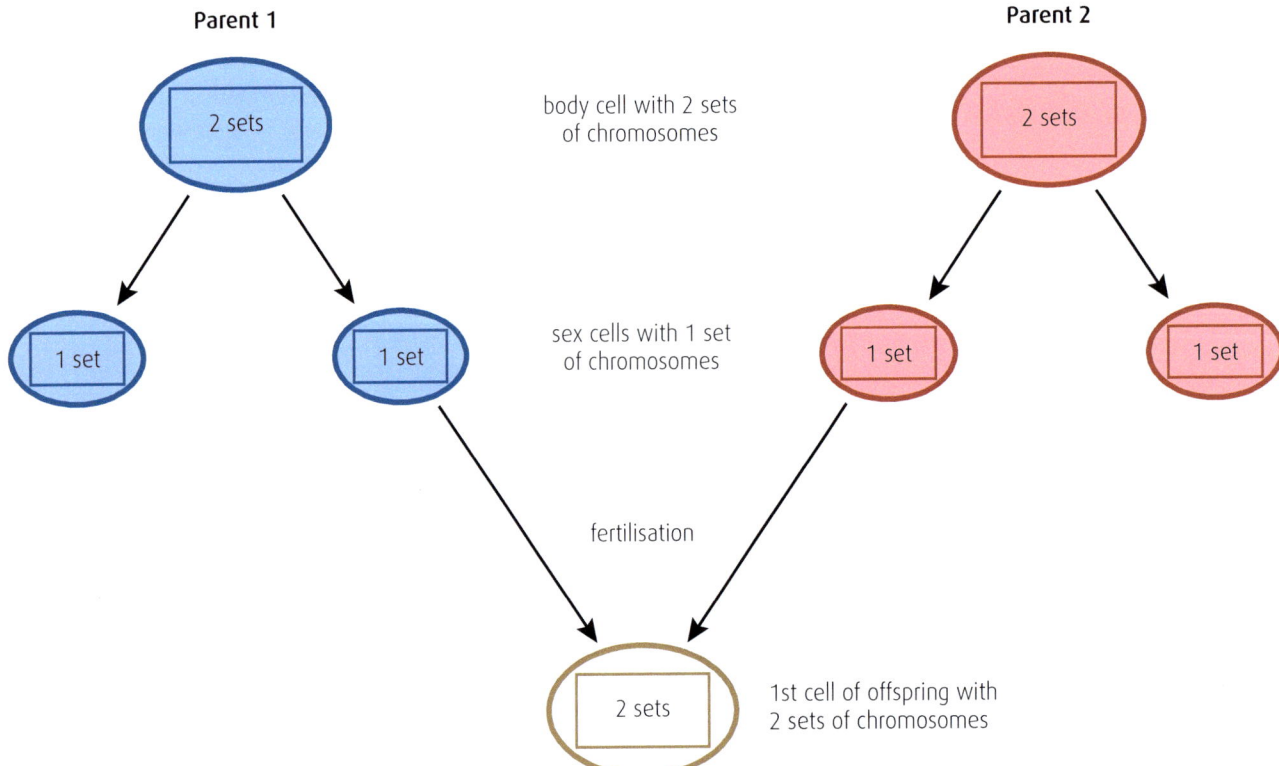

Features of sexual reproduction

ASEXUAL REPRODUCTION

Not all organisms are able to carry out asexual reproduction. It is common in single-celled organisms and other simple living things. It also takes place in many plants, including flowering plants.

Asexual reproduction basically involves an organism producing an outgrowth which becomes detached. The separated structure then develops into a complete new individual.

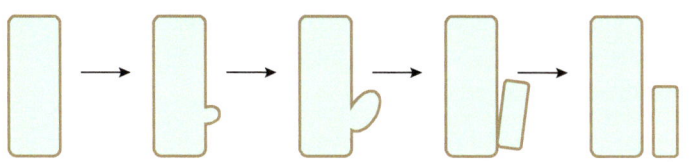

parent organism develops an outgrowth which separates to form a new individual

Features of asexual reproduction

DON'T FORGET

Sexual reproduction always involves the formation of sex cells with half the normal chromosome number. When two sex cells join at fertilisation, the normal chromosome number is restored. This does not happen in asexual reproduction.

ADVANTAGES OF SEXUAL AND ASEXUAL REPRODUCTION

ADVANTAGES OF SEXUAL REPRODUCTION

During sexual reproduction, an individual produces specialised sex cells which contain only one set of chromosomes. Other body cells contain two sets of chromosomes.

These sex cells show differences in their genetic information which produce variation in the offspring (see pages 54–55).

This is beneficial because it allows a species to adapt to changing environments and to become better adapted to the existing environment. Variation and adaptation are essential factors in the mechanism of evolution.

ADVANTAGES OF ASEXUAL REPRODUCTION

During asexual reproduction part of an organism becomes detached and develops as a separate individual. All of the cells which make up this new individual have been produced by normal cell division in the parent organism. This means that the genetic information in the new individual will be the same as that of the parent organism. It will also be genetically identical to any other offspring that are produced asexually by the parent.

Asexual reproduction does not have the advantages associated with sexual reproduction. However, all offspring will be just as well adapted to the environment as each other and as the parent. This method of reproduction allows rapid population growth in suitable conditions. It also has the benefit of not involving vulnerable embryo stages in the development of offspring.

SUMMARY

- Sexual reproduction involves the production of sex cells containing one set of chromosomes. A sex cell from each of two parent organisms combine at fertilisation to produce a cell with two sets of chromosomes. This is the first cell of the offspring.
- Sexual reproduction produces variation in the offspring.
- Asexual reproduction involves part of a parent organism becoming detached and developing as a new individual.
- Offspring of asexual reproduction are genetically identical to their parent.

THINGS TO DO AND THINK ABOUT

1. What type of cells contain only one set of chromosomes, rather than the usual two sets?
2. Explain the meaning of the term 'asexual'.
3. Describe the two essential features of sexual reproduction.
4. Explain why being genetically identical may be an advantage to offspring produced asexually.

SEXUAL AND ASEXUAL REPRODUCTION

PRODUCTION OF SEX CELLS IN FLOWERING PLANTS AND MAMMALS

FLOWERING PLANTS

The flower of a plant normally produces both male and female sex cells. The male sex cells are called pollen and are produced in structures called anthers. The female sex cells are called ovules and are produced in structures called ovaries.

Pollen must be carried from an anther to a structure called the stigma which is attached to the ovary. This process is called pollination. From the stigma the pollen nucleus can make its way to an ovule to fertilise it. The fertilised ovule develops into a seed which will eventually grow into a new plant.

Some plants depend on the wind for pollination and some depend on insects. The flowers of each type of plant may look very different but they contain the same structures.

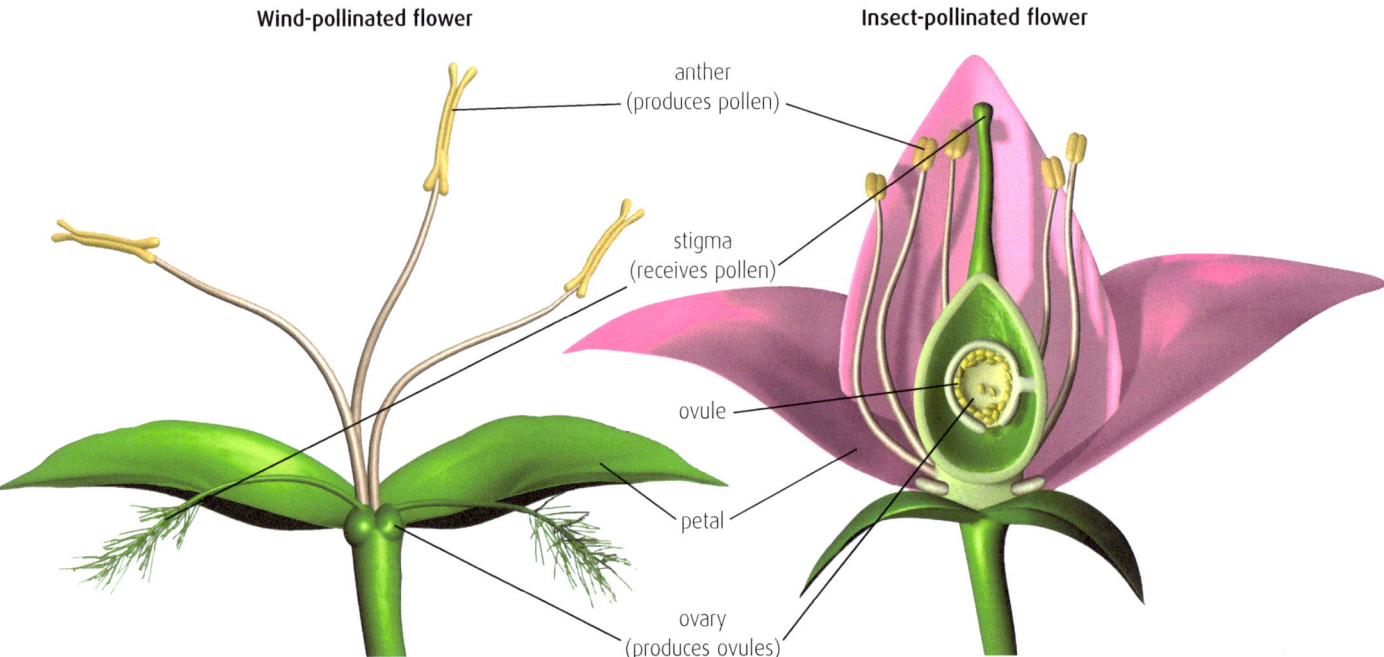

WIND POLLINATION

Features of wind pollination:

- small inconspicuous petals (no need to attract insects)
- anthers and stigmas exposed so wind can easily blow pollen
- pollen is small and light so easily carried by the wind
- large quantities of pollen produced to increase the chance of some reaching the stigma

Wind pollination is a hit-or-miss process. Most of the pollen released will never reach a suitable stigma and so will be wasted. An advantage of wind pollination is that the plants are not dependent on insects and do not have to compete with each other to attract them.

INSECT POLLINATION

Features of insect pollination:

- bright petals to attract insects
- produce nectar to attract insects
- anthers and stigmas enclosed so insects must brush past them

Insect pollination is less wasteful although some of the pollen and the nectar is used as food by insects. Pollen which becomes stuck to an insect's body has a good chance of being carried to another flower of the same kind and sticking to the stigma.

Sexual and Asexual Reproduction – Production of Sex Cells in Flowering Plants and Mammals

MAMMALS

Mammals possess reproductive organs which produce the sex cells. The male sex cells of animals are called sperm and they are produced in organs called testes (singular – testis). Female sex cells are called ova or eggs and are produced in ovaries.

In mammals sperm leave the testes and pass through the penis which can deposit the sperm in the vagina of a female. From here the sperm must swim towards the ovaries. If an ovum has been released from the ovary it may be fertilised by a sperm cell. If this happens the fertilised egg will become an embryo which will develop inside the uterus of the female.

male

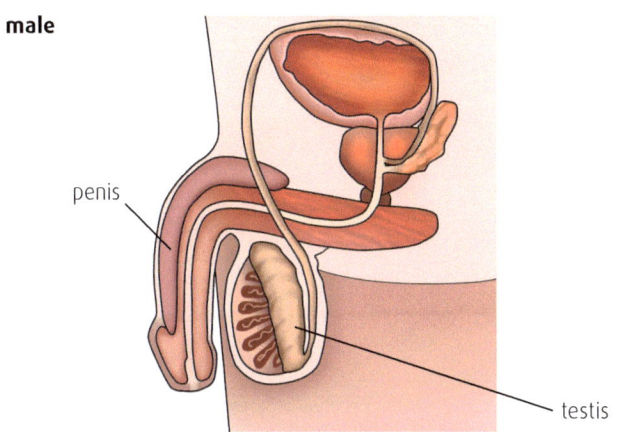

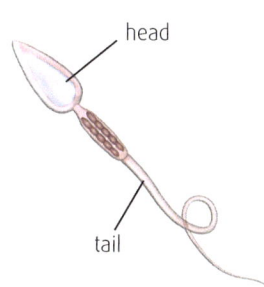

sperm cell

The head contains the nucleus with its chromosomes.

The tail allows the sperm to swim

female

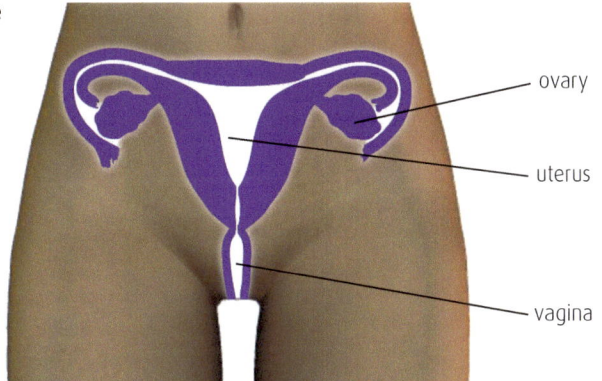

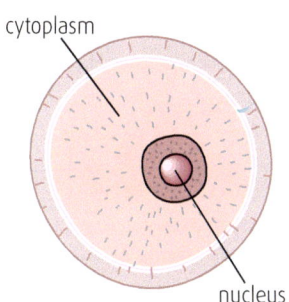

ovum or egg cell

The cytoplasm contains a food store and the nucleus contains the chromosomes

Human reproductive systems

SUMMARY
- The male sex cells of flowering plants are called pollen. They are produced by anthers.
- The female sex cells are called ovules and are produced by ovaries.
- The male sex cells of mammals are called sperm. They are produced by testes.
- The female sex cells are called ova or eggs and are produced by ovaries.

DON'T FORGET

In both mammals and flowering plants, the male sex cells are small and must move to the female sex cells. The female sex cells are large and contain a food store.

THINGS TO DO AND THINK ABOUT

1. Name the male and female sex cells of mammals and of flowering plants and the structures which produce them.
2. How do wind-pollinated plants overcome the fact that the chance of pollen achieving pollination is very low?
3. In mammals, how do sperm cells reach an egg cell? How are they adapted to do this?
4. What is contained in the nucleus of a sex cell?

SEXUAL AND ASEXUAL REPRODUCTION

SUCCESS RATES OF SEXUAL REPRODUCTION & METHODS OF ASEXUAL REPRODUCTION

SUCCESS RATES OF DIFFERENT SEXUAL REPRODUCTION METHODS

There are many different methods used by organisms for sexual reproduction. Different methods have different chances of successful fertilisation and offspring survival. Fertilisation may be internal (inside the body of the female) or external (in surrounding water). Internal fertilisation means that nutritious eggs are not eaten by predators. It also increases the chances of egg and sperm meeting for fertilisation.

If fertilisation is internal, development of the embryo may be internal or external. Internal development provides protection during critical early stages and so increases survival chances.

The amount of parental care shown by different organisms varies and this also affects survival chances.

The table below shows the characteristic features of vertebrate reproduction. The information refers to a species from each vertebrate group which produces a large number of eggs.

> **DON'T FORGET**
>
> A species whose offspring have a high chance of survival needs to produce fewer eggs than a species whose offspring have a lower chance of survival. This means that the parent does not waste body resources.

Vertebrate group	Fertilisation	Embryo development	Parental protection	Approximate number of eggs produced per year	Offspring surviving to breed	Success rate (%)
Fish	external	external	none	6 000 000	20	0·00033
Amphibians	external	external	protective jelly	20 000	40	0·2
Reptiles	internal	external	eggs buried	1000	10	1
Birds	internal	external	fed and protected by adult	50	2	4
Mammals	internal	internal	fed and protected by adult	50	4	8

METHODS OF ASEXUAL REPRODUCTION

There are many different methods by which organisms produce outgrowths which become new individuals. All of them result in offspring that are genetically identical to the parent organism.

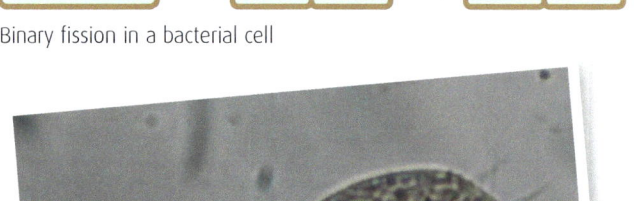
Binary fission in a bacterial cell

BINARY FISSION

This is cell division of a single-celled organism. It involves one cell dividing into two cells and the two cells separating to form two individuals.

Binary fission is shown in the diagram and photographs.

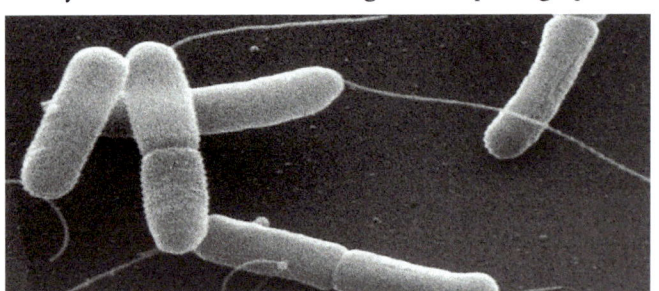
E. coli

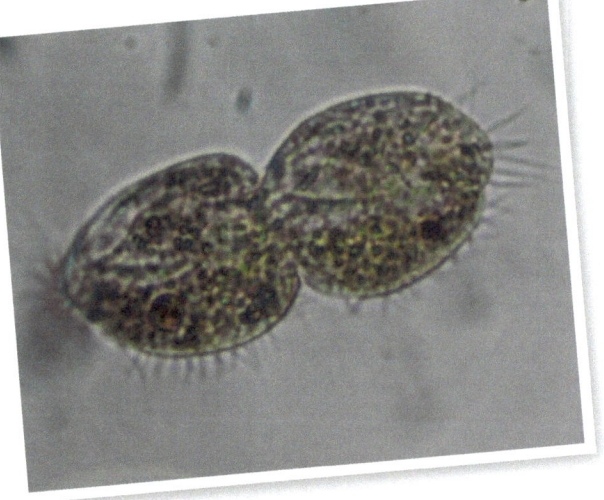
Paramecium

BUDDING

This is when there is an unequal division of the parent organism to produce a small outgrowth which becomes independent. It takes place in the single-celled fungus, yeast. It also occurs in some simple multicelled organisms such as Hydra.

VEGETATIVE PROPAGATION

This term covers various methods of asexual reproduction which take place in some flowering plants. Some of these are dealt with in the next section.

SPORES

Spores are small cells produced by fungi and some algae. The spores are dispersed from the parent and develop into new organisms.

Mushrooms, toadstools and moulds are the spore-forming structures of fungi. In each case, most of the fungus is not visible and consists of a mass of microscopic strands which grow under the soil or through the tissues of other organisms.

FRAGMENTATION

Fragmentation is the development of a complete individual from a fragment of the parent organism. It occurs in animals, plants and fungi. It can be an extreme form of regeneration (see page 8) where fragments of an organism are broken off and fully regenerate into complete organisms. In other cases, fragmentation takes place as a deliberate method of reproduction.

Examples of budding

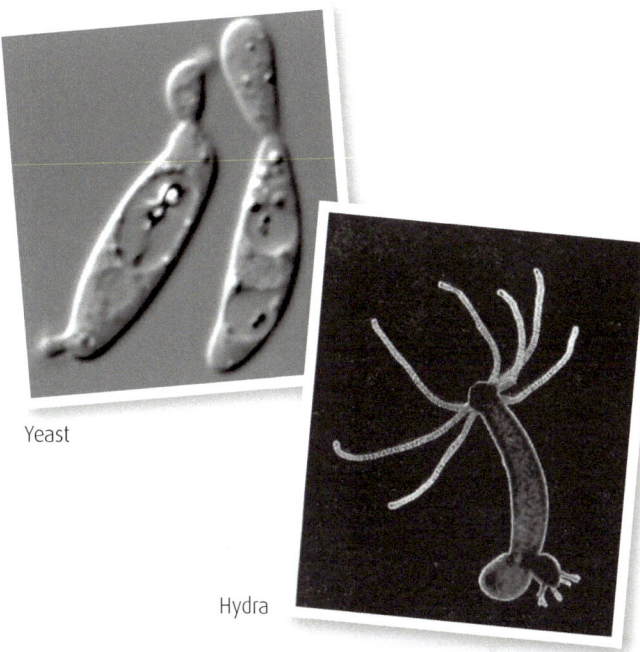

Yeast

Hydra

Clouds of spores being released by a Puffball mushroom

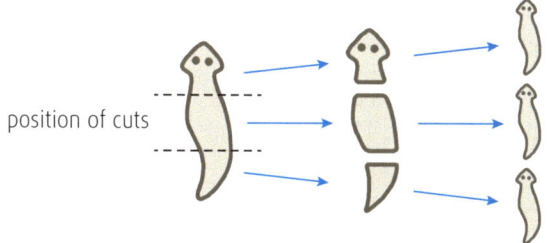

position of cuts

Fragmentation in Flatworm

SUMMARY

- The chances of successful reproduction are increased for organisms which use internal fertilisation of the eggs, internal development of embryos and extended parental care of the young.
- Many organisms are able to reproduce asexually to produce offspring that are genetically identical to the parent. This can be useful because all offspring have an equal chance of survival in a suitable environment.

THINGS TO DO AND THINK ABOUT

1. Explain why external fertilisation is not possible for land-living organisms.
2. Explain why the survival chances of the offspring of birds is greater than those of reptiles.
3. A frog produced 400 eggs in one year. 175 of the eggs failed to develop, 190 of the tadpoles were eaten by fish and 15 of the young frogs were eaten before they developed fully. The remainder all survived to breed.
   (a) How many of the original eggs developed into surviving frogs?
   (b) Calculate the percentage survival of the eggs.
4. Name the method of asexual reproduction which most closely resembles:
   (a) normal cell division
   (b) regeneration.

PROPAGATING AND GROWING PLANTS

SEXUAL AND ASEXUAL REPRODUCTION

Plant propagation (producing new plants) can take place in a number of ways.

All flowering plants can reproduce sexually by producing seeds as the result of pollination and fertilisation. For some plants, this is their only way of reproducing. This can be a disadvantage for plant growers because sexual reproduction always results in variation in the offspring. This means that plants grown from seed will differ in some ways from the parent plants and from each other.

ASEXUAL REPRODUCTION

Some plants can reproduce asexually. This is an advantage for plant growers because the offspring will be genetically identical to the parent. Therefore the characteristics of the offspring can be guaranteed. This applies to crop plants such as potatoes, strawberries and apples. It also applies to ornamental plants such as daffodils, tulips and roses.

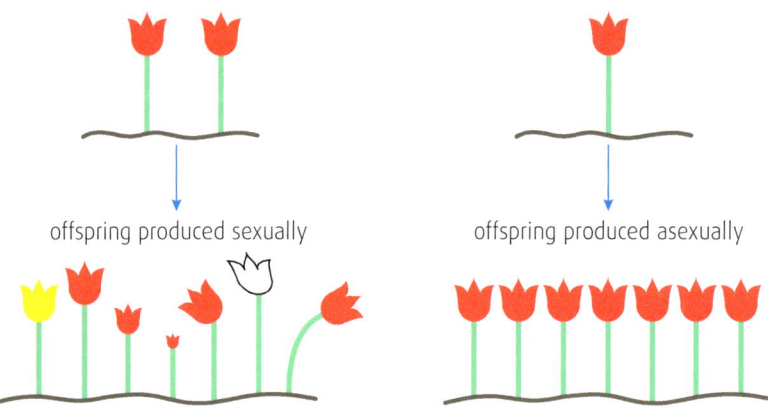

Sexual and asexual propagation

SEXUAL REPRODUCTION

Pollination results in pollen grains being transferred from the anthers of a flower to the stigma of another flower of the same plant species. It does not matter whether pollination is by wind or insects; the following processes of fertilisation and seed formation are the same.

A pollen grain produces a tube which grows down through the stigma into the ovary until it meets an ovule. The nucleus of the pollen passes down the tube and enters the ovule. Here it joins with the ovule nucleus. The joining of the two nuclei is called fertilisation.

Each fertilised ovule then develops into a seed which contains an embryo plant and a food store. The ovary, which now contains seeds, then develops into the fruit of the plant. The fruit helps to spread the seeds away from the parent plant. Seed dispersal helps to prevent competition for light, water and nutrients between the parent plant and germinating seeds.

> **DON'T FORGET**
>
> Sexual reproduction always leads to variation in the offspring. Asexual reproduction produces offspring that are genetically identical to the parent.

SEED FORMATION

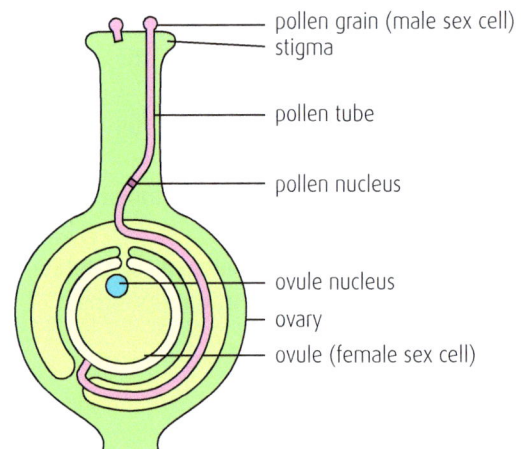

Growth of the pollen tube

SEED DISPERSAL

Seeds can be dispersed away from the parent plant in a number of different ways. Wind and animals are the most common methods of dispersal. The changes to the ovary, which take place after fertilisation, help to make this possible.

These seeds are very light with fine hairs which act as a parachute, or else have thin extensions which act like wings.

Examples of seeds dispersed by the wind

Dandelion seeds

Sycamore seeds

Propagating and Growing Plants – Sexual and Asexual Reproduction

These seeds have hooks which cling to the coats of animals.

Examples of seeds dispersed by animals on the outside of their bodies

Burdock seeds

Cleavers or 'Sticky Willy' seeds

These seeds are found either inside or on the outside of juicy fruits. Animals eat the fruit and the seeds pass out in the animals faeces, unharmed.

Examples of seeds which are dispersed by animals inside their bodies

Hawthorn berries

Strawberry

SUMMARY

- All flowering plants can reproduce sexually by producing seeds. Some flowering plants can also reproduce asexually.
- Plants which grow from seeds show differences to each other and the parent plants.
- Plants which are produced asexually are genetically identical to the parent.
- Sexual reproduction in plants involves a number of stages:
  - pollination – the transfer of pollen from anther to stigma
  - growth of a pollen tube – to allow the male (pollen) nucleus to reach the female (ovule) nucleus
  - fertilisation – the joining of the pollen nucleus with the ovule nucleus
  - seed formation – containing an embryo plant and a food store
  - seed dispersal – to avoid competition with the parent plant.

THINGS TO DO AND THINK ABOUT

1. The seeds from the inside of an apple were collected and planted. All of the seeds grew into trees. Would these trees produce apples of the same variety as the one they came from? Explain your answer.
2. Describe two similarities between the male sex cells of flowering plants and those of mammals. Do the same for the female sex cells.
3. Describe the process which must take place after pollination before seeds can be formed.
4. Explain why it is beneficial for seeds to be dispersed away from the parent plant.

PROPAGATING AND GROWING PLANTS

ASEXUAL PROPAGATION: NATURAL METHODS

Asexual propagation is used to produce more plants with known characteristics. This is because all the offspring are genetically identical to the parent plant. Some of the methods are natural and occur as a normal means of reproduction of the plant. Other methods are artificial and make use of a plant's ability for regeneration.

The following different methods allow plants in nature to reproduce and which are used by plant growers to produce more plants.

BULBS

A bulb is a structure which contains swollen leaves. The swollen parts contain a food store which allows the plant to survive underground during the winter while the rest of the plant dies away. One plant may produce several bulbs. Each of these can grow into a new plant the following year and so the number of plants increases year by year.

Bulbs are produced by onions, daffodils, tulips and many other species of plants.

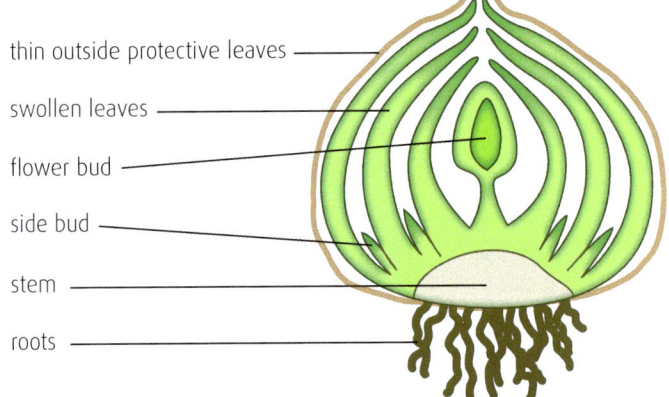

Section through a bulb

CORMS

A corm is similar to a bulb in that it allows a plant to survive over winter and increase in number. The difference is that it is the base of the plant stem which contains the stored food and not leaves. One plant may produce several corms, each of which can produce a new plant the following year.

Corms are produced by crocuses, gladioli, bananas and many other plant species.

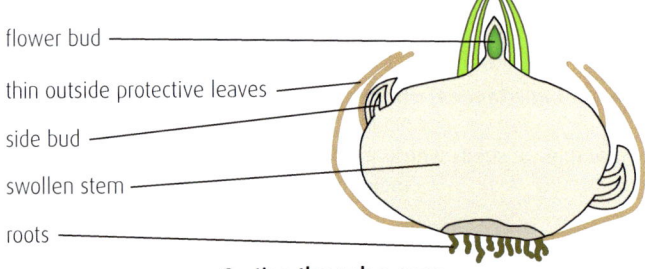

Section through a corm

TUBERS

Tubers are also parts of plants which become swollen with stored food to allow the plant to survive over winter. One plant can produce several tubers, each of which can grow into a new plant the following year. In some species tubers may form as swollen parts of underground stems. In other species tubers are formed as swollen parts of roots.

Plants which produce stem tubers include potatoes, cyclamen and some begonias.

Plants which produce root tubers include dahlias, yams and sweet potatoes.

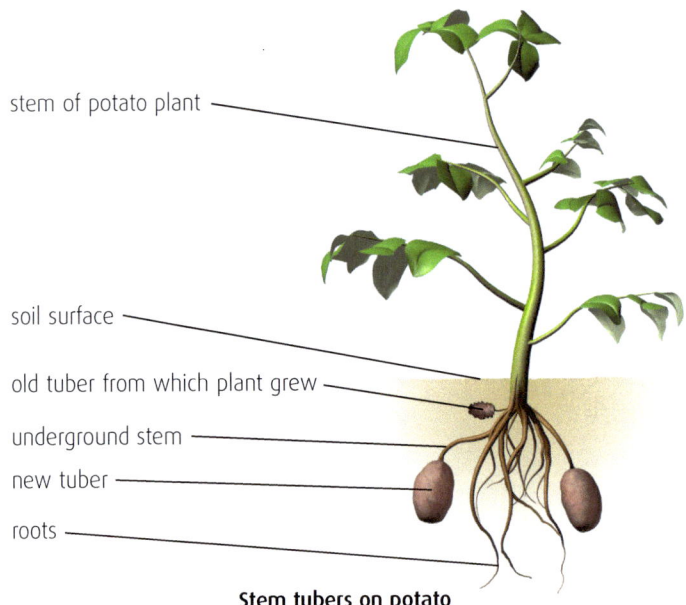

Stem tubers on potato

RHIZOMES

Rhizomes are horizontal plant stems which usually grow and spread just below the soil surface. Vertical shoots grow upwards to form new leafy plants. Pieces of rhizome which have broken off from the main plant can grow into new plants.

Rhizomes also allow plants to survive over winter.

Plants which produce rhizomes include bamboo, hops, ginger, irises and many grasses.

Rhizome of bamboo

RUNNERS OR STOLONS

These are similar to rhizomes. They are horizontal stems which usually grow on the soil surface. Buds along the runner produce roots and leaves, forming a new plant.

Other plants which produce runners include spider plant, mint and some grasses.

Runners of strawberry

SUMMARY
- Asexual propagation techniques allow the plant growers to produce large numbers of plants that are genetically identical to a parent plant. This means that the characteristics of the new plants can be guaranteed.
- Bulbs, corms, tubers, rhizomes and runners are all natural methods of plant reproduction which plant growers use to increase the number of plants.

DON'T FORGET

All of the above methods of asexual propagation take place naturally but plant growers use them to increase the numbers of plants they wish to grow.

THINGS TO DO AND THINK ABOUT

1. Name two natural methods of propagating plants asexually.
2. Describe the procedure and the sequence of events which would allow a gardener to produce a lot of potatoes from a single potato tuber.
3. As well as increasing the number of plants, what other function do structures such as bulbs, corms and tubers have?
4. Bulbs and corms look very similar. What is the main difference between them?

PROPAGATING AND GROWING PLANTS

ASEXUAL PROPAGATION: ARTIFICIAL METHODS

Plant growers have developed artificial methods of producing more plants asexually. Some of these methods depend on the capability of the plant to regenerate.

CUTTINGS

A cutting is a part of a plant that has been cut off a parent plant and planted in soil. The cutting produces roots and develops into a complete plant. The cutting may be treated with a plant hormone to encourage the growth of the new roots. Cuttings are taken as sections of stem from some types of plants, or as leaves from other types.

1. Cut through shoot just below a leaf joint
2. Remove leaves from bottom part of cutting
3. Place cutting in well watered compost

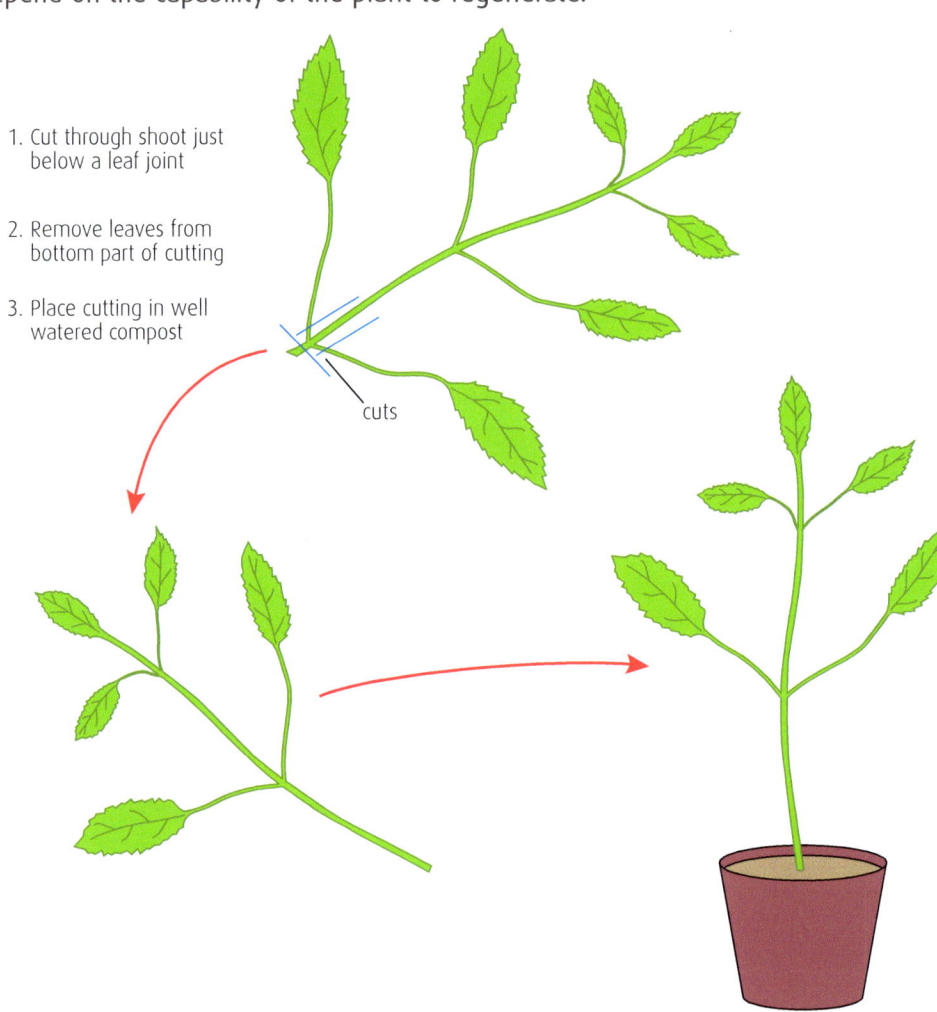

Taking a stem cutting

1. Cut through leaf stalk
2. Dip end of stalk into rooting powder
3. Place leaf into well-watered compost

Taking a leaf cutting

48

GRAFTING

A graft is a union between two plants. A stem of one plant (the scion) is joined to the roots and lower stem of another plant (the rootstock).

The rootstock influences the size and vigour of the growth of the scion but the scion will retain its own characteristics when it grows.

Grafting is used for the propagation of shrubs and trees which have desirable features such as types of flowers or fruit. It is used when the type of shrub or tree does not grow well or does not propagate easily from cuttings.

When the graft heals, the rootstock provides the scion with water and nutrients from the soil, allowing it to grow.

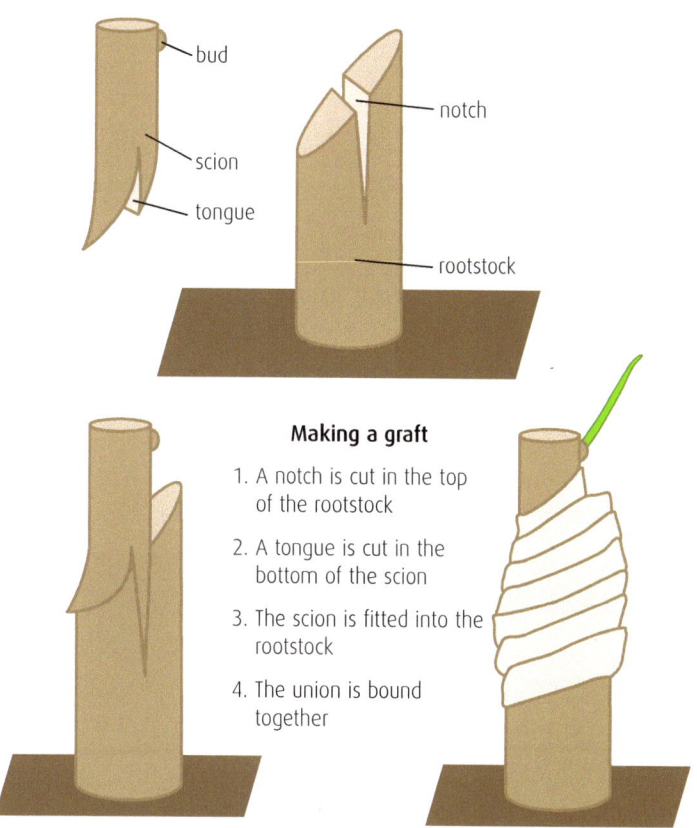

Making a graft

1. A notch is cut in the top of the rootstock
2. A tongue is cut in the bottom of the scion
3. The scion is fitted into the rootstock
4. The union is bound together

TISSUE CULTURE

Tissue culture is the growth of complete plants from a few isolated cells taken from a parent plant.

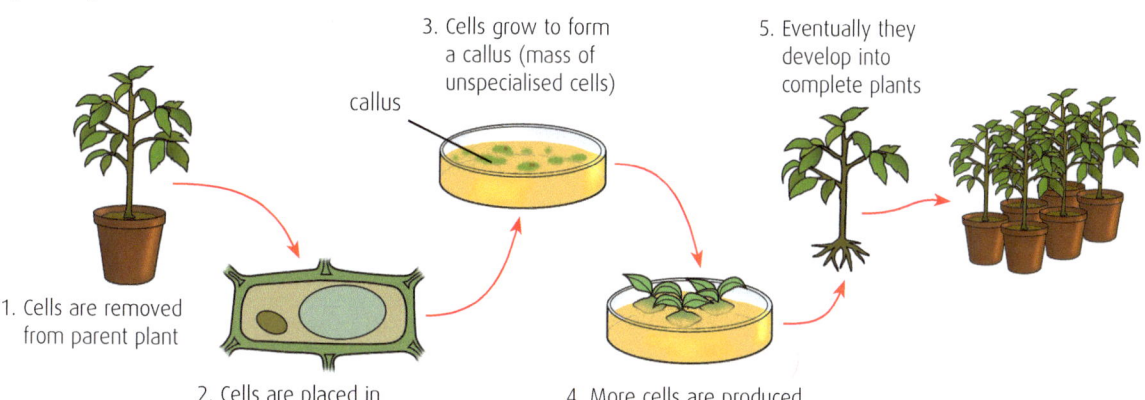

1. Cells are removed from parent plant
2. Cells are placed in containers with nutrients
3. Cells grow to form a callus (mass of unspecialised cells)
4. More cells are produced and become specialised
5. Eventually they develop into complete plants

SUMMARY

- Artificial methods of asexual plant propagation depend on the ability of plants to regenerate damaged parts.
- Rooting powder contains plant hormones which encourage the growth of new roots.
- Taking cuttings and making grafts are traditional propagation techniques.
- Tissue culture is a modern method of plant propagation.

DON'T FORGET

Asexual propagation techniques used with plants depend on the ability of plants to regenerate damaged parts.

THINGS TO DO AND THINK ABOUT

1. Name two artificial methods of propagating plants asexually.
2. What is the purpose of the plant hormone powder which is sometimes used when growing plants from cuttings?
3. After making a graft, does the new growth show the characteristics of the scion or the rootstock?
4. What is a callus?

COMMERCIAL USES OF PLANTS

FOOD & FUELS

Plants are of vital importance to the lives of other organisms. The process of photosynthesis (see pages 26–27) enables them to make food for themselves. This means that they are the producers of the food chains which support all other living things. Photosynthesis also produces the oxygen which is needed by other organisms for respiration (see pages 30–31).

Plants have always been used by people for food, fuel, medicine, timber, fabrics and for ornamental purposes. Selective breeding has improved the plants and made them better suited to these uses. This involves careful cross-breeding of particular plants which have desirable features. The hope is to produce offspring which show improvements in these features.

Today, plant scientists are developing biotechnological methods of using plants as living factories to produce a variety of chemical products.

FOOD

Plants are grown as food, either for consumption by humans or as food for livestock.

Food production has increased because of improvements in agriculture. This has been the reason for the rapid growth of the human population. These improvements include:

Glasshouse strawberry production

- Mechanisation – the increased use of machines to replace manual labour.
- Monoculture – the large-scale cultivation of a single crop. This is more efficient although it may encourage the spread of specific pests and diseases in the crop.
- Pesticides – the use of chemicals to kill other organisms which may cause damage to the crop. There are different types of pesticides:
Herbicides kill other plants which compete with the crop.
Insecticides kill insects which eat the crop.
Fungicides kill fungi which cause disease to the crop.
Pesticides can cause problems by affecting other organisms in the ecosystem. Pest species may develop resistance to the pesticide, making them less effective.
- Fertilisers – chemical nutrients which are needed by plants. Modern agriculture does not allow the natural replacement of soil nutrients through the decomposition of dead and waste material. To allow continued use of the soil to grow crops, nutrients must be replaced using artificial fertilisers.
Fertilisers can cause problems if they pass from cultivated fields into the wider environment.
- Irrigation – bringing water to dry areas to allow crop growth.
- Plant breeding – developing new varieties which grow better.
- Controlling the conditions of growth of crop plants – using large glasshouses, the light, temperature and availability of carbon dioxide, water and plant nutrients can be adjusted. This can increase growth rates.
- Genetic modification – inserting new genes into crop plants to produce increased yields. The modified plants may show improvements such as faster growth, improved resistance to pests or they may require less fertiliser.

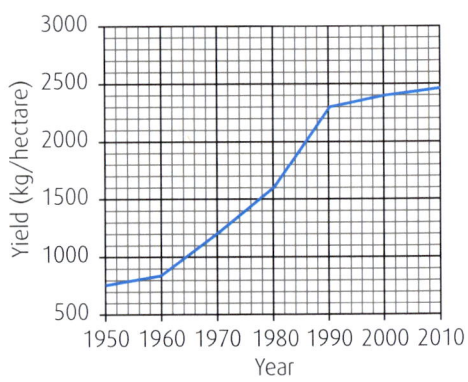

Increases in wheat yields of developing countries

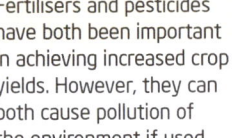

Fertilisers and pesticides have both been important in achieving increased crop yields. However, they can both cause pollution of the environment if used in excess.

Commercial Uses of Plants – Food & Fuels U2

FUELS

Wood has always been an important fuel for humans. In many parts of the world, wood is still the main fuel for domestic use.

Today, biofuels are being produced from plants as full or partial replacements for petrol and diesel, which come from fossil fuels.

- Crops such as sugar cane, sugar beet and maize are grown not only as foods but to provide sugar for fermentation by yeast. The alcohol that is formed is mixed with conventional petrol to power petrol engines.
- Crops such as rapeseed, sunflower and palm produce vegetable oils which can be used to make biodiesel for use in diesel engines.
- Waste vegetable matter can be treated in digesters to produce sugars for fermentation.
- Microscopic algae can be grown to produce biomass which is processed to produce fuels.

Biofuels will probably become commercially more important as reserves of fossil fuels decline.

Fuels produced from plant material are considered to be better for the environment than fossil fuels. Both types of fuel release carbon dioxide into the atmosphere when they burn. However, plant crops that are grown to produce fuels have removed an equivalent amount of carbon dioxide from the atmosphere during their growth. This means that they do not cause an increase in atmospheric carbon dioxide.

SUMMARY

- Increases in crop yields have been the main reason for the dramatic rise in the human population. Traditionally, plant breeding and improvements in cultivation have been the reasons for yield increases.
- Genetic modification can introduce new features into plants. This technique could become more widespread, although concerns still exist about its use.
- Plants are becoming increasingly important as a source of fuels to replace fossil fuels such as petrol and diesel.

THINGS TO DO AND THINK ABOUT

1. Name two uses of plants apart from food, fuel and medicine.
2. Give three ways of improving crop yields.
3. Explain why burning biofuels is not considered to cause an increase in the carbon dioxide concentration of the atmosphere.
4. There are ethical arguments about growing crops for fuel rather than as food. Give one example of biofuel production which does not involve this choice.

COMMERCIAL USES OF PLANTS

MEDICINES

People have always used plants for traditional treatments of illness and injury. There is a large industry based on herbal remedies. Herbalism is generally considered to be a form of alternative medicine which is not based on scientific development and testing. However, there are many conventional medicines which have been produced from plants. Some of these medicines can be synthetically produced.

It is thought that many potential new medicines have still to be discovered in plant species that have not yet been fully studied. It is believed that the large-scale destruction of habitats such as tropical rainforests will cause the extinction of many plant species before their medicinal value is known.

The table contains information about several medicines derived from plants.

Plant	Medicine	Use
Opium poppy	Morphine	Pain relief
Foxglove	Digitalis	Treatment of heart disease
English yew	Taxols	Treatment of breast cancer
Cinchona tree	Quinine	Treatment of malaria
Willow	Salicylic acid (aspirin)	Pain relief

Modern biotechnological techniques are likely to increase the importance of plants in the production of medicines.

PHARMING

Pharming (see page 35) is the use of genetic engineering to insert genes that code for useful protein pharmaceuticals from one organism into different animals or plants. This creates a genetically modified (GM) organism that is able to produce a substance which otherwise it would be unable to do. One method of inserting the new gene into a plant is to add the gene to the DNA of a virus. The plant is then infected with the modified virus.

Pharming may use whole plants which are harvested and then the desired protein is extracted from them. Techniques are being developed to use tissue cultures rather than whole plants to produce the protein. This would make the extraction of the protein easier.

There are advantages of using plants for pharming.

- Plants do not carry diseases which affect humans.
- Any useful substance produced in the fruits or seeds of a plant will be able to be stored using the same methods used to store the fruits or seeds.
- Plants are able to manufacture more complex substances than microorganisms.
- Plants are also likely to be cheaper than using animal cell cultures for the production of such substances.
- Pharmaceutical crops could be useful in tackling disease in developing countries. They could be used to produce vaccines more cheaply and help prevent diseases such as measles and hepatitis.

Pharming is a form of genetic engineering and therefore there are general arguments against it (see page 35).

- Although pharmed plants are not grown as food crops, there is a concern that the altered plants might find their way into the food supply or cross-pollinate with conventional crops. Such incidents have happened with genetically modified crops in the past.
- There are concerns that some of the chemicals produced by pharmed plants may cause allergies.

Commercial Uses of Plants – Medicines U2

The moss bioreactor is used to produce proteins from genetically modified moss plants.

Moss bioreactor

SUMMARY

- Plants are commercially important because they have an increasing range of uses.
- Pharming allows plants to be used to produce specific protein compounds which have a medicinal value. A gene which codes for the protein is inserted into the plant, giving it the ability to manufacture that protein.

DON'T FORGET

Plants have been the source of many traditional medicines for thousands of years.
There are probably many new medicines still to be discovered in plants.

THINGS TO DO AND THINK ABOUT

1. How does herbalism differ from conventional medicine in the use of plants for medicinal purposes?

2. Name two conventional medicines that have been discovered in plants, and give their uses.

3. Describe one method by which a new gene may be inserted into a plant.

4. Why are genetically modified plants being considered to produce new medicinal substances, rather than microorganisms?

GENETIC INFORMATION
SPECIES & VARIATION

SPECIES

A species is a group of organisms which are able to interbreed and produce fertile offspring.

Sometimes individuals from closely related species can mate and produce hybrid offspring, but these offspring are usually infertile and cannot produce more offspring like themselves. This can occur in the wild, for example between polar bears and North American brown bears, but such unusual hybrids are normally the result of mating in artificial conditions.

Examples of hybrids

DON'T FORGET

Just because two animals can mate and produce an offspring, it does not mean they are members of the same species.

Zorse (zebra and horse)

Liger (lion and tiger)

VARIATION

Although individuals of the same species share genetic information for the same range of characteristics, they are not identical. This is because a characteristic can have more than one form due to small differences in the gene, or genetic information, which codes for that characteristic. The differences in a characteristic, which occur between individuals, are called variations.

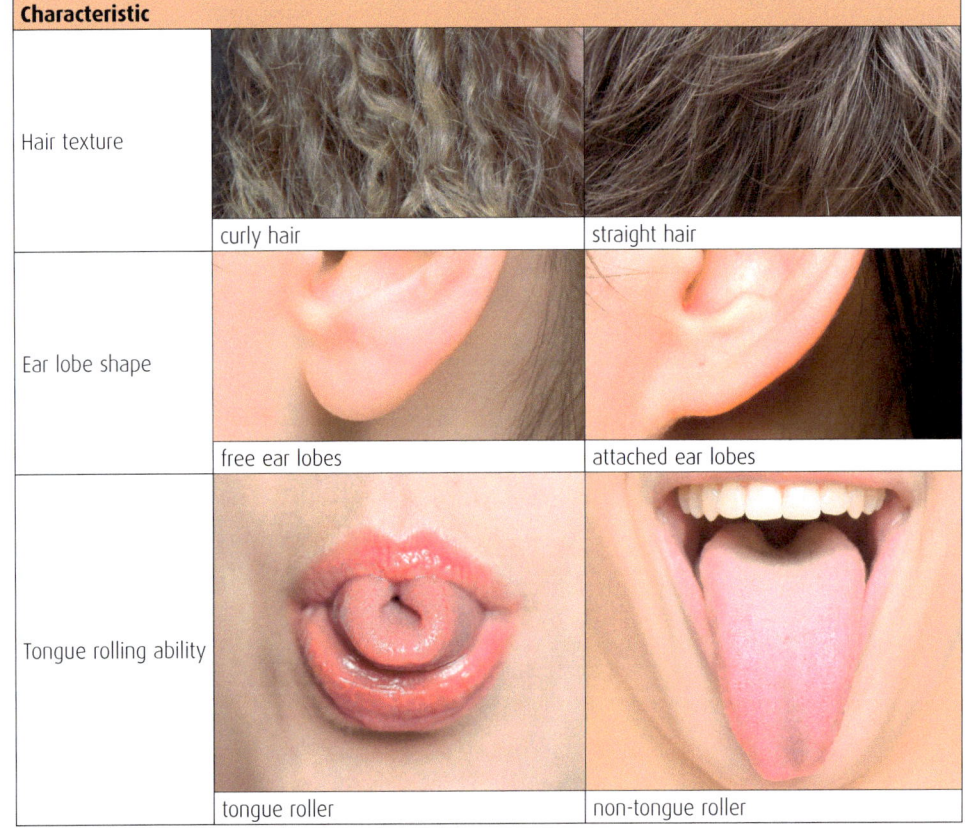

Characteristic		
Hair texture	curly hair	straight hair
Ear lobe shape	free ear lobes	attached ear lobes
Tongue rolling ability	tongue roller	non-tongue roller

Examples of human variation

Genetic Information – Species & Variation U2

PHENOTYPE AND GENOTYPE

Phenotype – this is how a characteristic appears in an individual. For the tongue-rolling characteristic there are two possible phenotypes: tongue roller or non-tongue roller.

Genotype – this is the genetic information, or genes, which an individual possesses for a particular characteristic.

Allele – this is a form of a gene. There are two versions, or alleles, of the gene for the tongue-rolling characteristic:

- The allele which gives the ability to tongue roll can be represented by **T**.
- The allele which does not give the ability can be represented by **t**.

Each body cell of an organism has two matching sets of chromosomes which carry the genes. Therefore there are two alleles present which make up the genotype for each characteristic.

This means that for the tongue-rolling characteristic, there must be three possible genotypes. These are: **TT tt Tt**

However, there are only two phenotypes because one allele shows its effect over the other if both are present. **T** shows its effect over **t**. **T** is said to be dominant and **t** is said to be recessive.

The result of this is shown in the table.

> **DON'T FORGET**
>
> In body cells, chromosomes are present in pairs. Therefore there are two alleles which make up the genotype for a characteristic.
> The dominant allele is normally shown as an upper case letter. The recessive allele is shown as the same lower case letter.

Phenotype		Genotype
Tongue roller		**TT** or **Tt**
Non-tongue roller		**tt**

SUMMARY

- Members of a species can interbreed to produce fertile offspring. Hybrids (offspring from individuals of closely related species) are usually infertile.
- Variations are differences in the characteristics of individuals. They are caused by different alleles of genes. Usually one allele is dominant and will show its effect over other alleles of the same gene.
- Phenotype means the appearance of a characteristic in an individual. Genotype means the combination of alleles which causes the characteristic.

THINGS TO DO AND THINK ABOUT

1. What is meant by the term 'hybrid'?
2. Why is it not possible for a population of Zorses containing several generations to exist?
3. What is the term used to describe the genetic information that an individual possesses for a characteristic?
4. **P** is the allele for the ability to taste the chemical PTC.
   p is the allele for not being able to taste PTC.
   F is the allele for having freckles.
   f is the allele for not having freckles.
   W is the allele for having a widow's peak hairline.
   w is the allele for not having a widow's peak hairline.
   Give the phenotypes for each of the following genotypes.
   pp   PP   Ff   ff   Ww   ww

GENETIC INFORMATION

INHERITANCE

Inheritance is the passing of genetic information from parent to offspring. Sometimes it is easy to see features which have been inherited.

Two generations of males

THE MECHANISM OF INHERITANCE

The body cells of an organism contain two sets of chromosomes and therefore two alleles for each characteristic. The sex cells contain only one set of chromosome and so they have only one allele of each gene. If two different alleles of a gene are present in body cells, 50% of the sex cells will contain one type of allele. The other 50% of the sex cells will contain the other type of allele.

THE INHERITANCE OF RED HAIR

Red hair is caused by a particular gene which has two alleles, R and r. R is dominant and r is recessive.

For a person to have red hair, they must possess two r alleles. Their genotype for this characteristic is rr.

If they possess either of the other possible genotypes, RR or Rr, then they will not have red hair. Their hair colour will then depend on other genes.

Neither of the two people represented here has red hair because each has the Rr genotype. When they produce sex cells (sperm or eggs), half will contain the R allele and the other half will contain the r allele.

Two generations of females

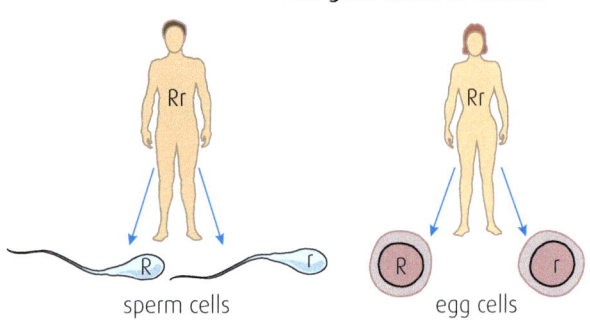

Inheritance of red hair

If these people have children together, there are different combinations of the R and r alleles possible at fertilisation.

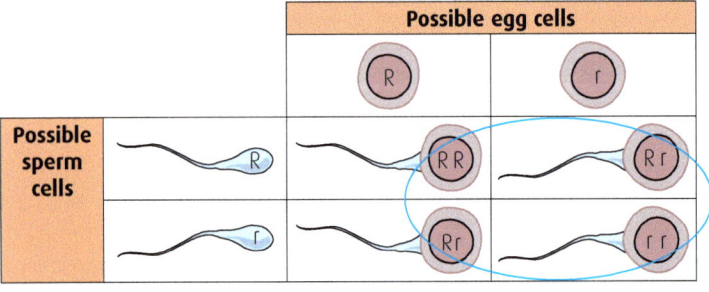

DON'T FORGET

Body cells carry two alleles of every gene but sex cells carry only one allele.

Possible genotypes of children from this couple

This shows that out of the four possible combinations of egg and sperm cells at fertilisation, only one will give the rr genotype needed to produce red hair. Therefore it would be expected that one quarter of the children of this couple would have red hair, even though neither parent had red hair.

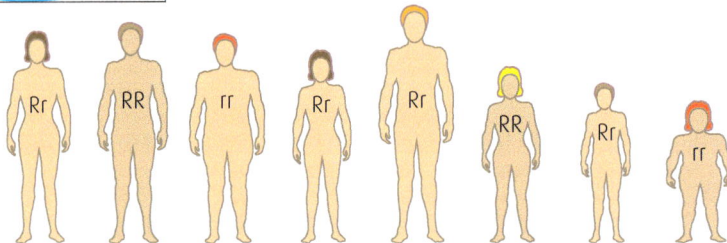

Expected genotypes and phenotypes of children

Several human characteristics are thought to be controlled by a single gene with different alleles, in the same way as red hair colour. However, most characteristics are influenced by many different genes acting together. This produces a much more complex pattern of inheritance.

Genetic Information – Inheritance   U2

INHERITANCE OF GENDER

The gender, or sex, of a person is an inherited characteristic. However, the mechanism for the inheritance of sex is different from other inherited characteristics. It is not controlled by a single gene. Instead it is controlled by a pair of chromosomes. Humans have 23 pairs of chromosomes in their body cells and one pair of these are the sex chromosomes. If all the chromosomes of a male and female are photographed and arranged in order of their size, the difference between them can be seen.

Male chromosome pairs

There are 22 pairs of chromosomes which are similar in both sexes. In addition, females have another pair of matching chromosomes called X chromosomes. In males, this other pair of chromosomes do not match. One of them is an X chromosome and the other is a Y chromosome. These are the sex chromosomes.

When sex cells are produced, only one sex chromosome will be present in each cell. The egg cells from a female will always contain one X chromosome because that is the only type of sex chromosome she has. Half of the sperm cells a male produces will contain an X chromosome and the other half will contain a Y chromosome.

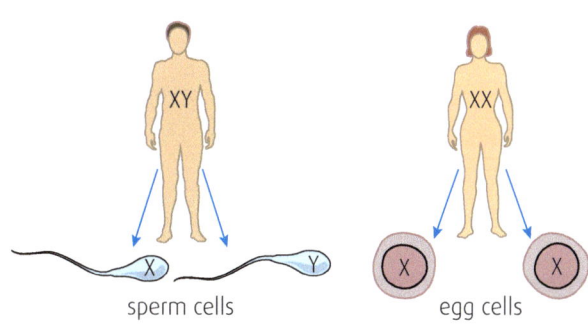

Inheritance of gender

At fertilisation, there are different possible combinations of sperm and eggs.

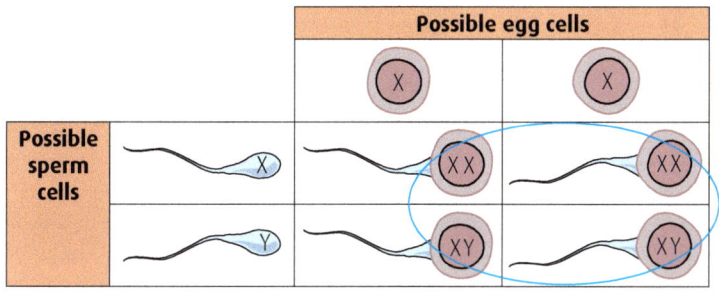

Possible genotypes of children from this couple

DON'T FORGET

Fertilisation is random. It is not certain that eggs and sperm will meet to give the expected results. The expected results are more likely to be seen when large numbers of offspring are produced, than with small numbers.

DON'T FORGET

Sex is determined by the inheritance of a pair of chromosomes, not by a particular gene.

This shows that there is an equal chance of an egg cell being fertilised by a sperm cell carrying an X chromosome as by one carrying a Y chromosome. It means that half the children in a family should be boys and half should be girls. As always, the random nature of fertilisation means that the expected proportions are not always produced.

SUMMARY

- Organisms possess two alleles for each inherited characteristic. The alleles can code for different variations of the characteristic. If the two alleles are different, usually one will be dominant and show its effect. The other allele is recessive and will not show its effect but it can still be passed on to the next generation.
- The sex of a person is determined by a pair of chromosomes called the sex chromosomes. Males possess one X and one Y chromosome. Females possess two X chromosomes.

THINGS TO DO AND THINK ABOUT

1. What term is used for different forms of a gene?
2. In the inheritance of red hair, which allele is dominant?
3. What is the chance of any child from each of the couples described below having red hair?
   Father Rr × Mother Rr
   Father RR × Mother rr
   Father rr × Mother RR
4. Explain why the sex of a child is determined by its father.

57

GROWTH AND DEVELOPMENT IN DIFFERENT ORGANISMS

GROWTH CURVES, GROWTH IN PLANTS & GROWTH IN ANIMALS

GROWTH CURVES

Growth is an increase in the size and mass of an organism. It is caused by cell division (which increases the number of cells in an organism) and cell growth (which increases the size of the cells).

Most organisms show a similar pattern of growth over their lifetime. If the size of an organism is measured over its lifetime a typical S-shaped growth curve is produced.

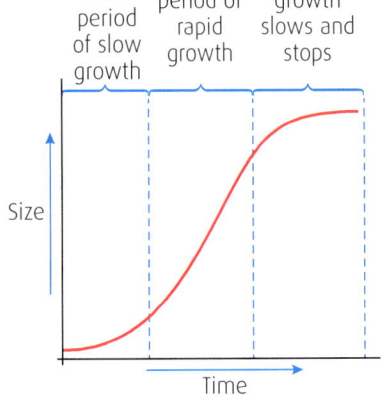

Typical growth curve

GROWTH IN PLANTS

Plants have localised areas of growth at the tips of roots and shoots and within the roots and shoots. In these areas, cells retain the ability to divide.

- Growth areas at the tips allow the plant to increase in length.
- Growth areas within the stems and roots allow them to increase in width.

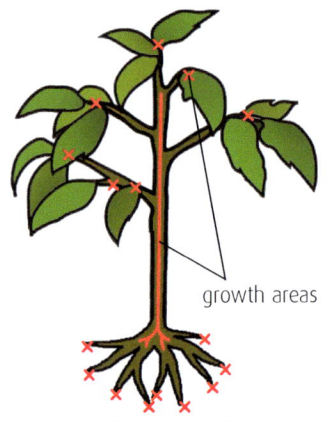

Growth areas in plants

GROWTH CURVES IN PLANTS

EXAMPLE 1   Annual plants

These grow from a seed, then flower and die in one year.

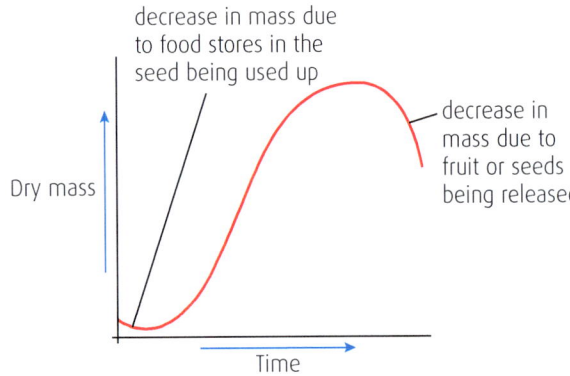

Annual plant growth curve

EXAMPLE 2   Deciduous trees

These trees lose their leaves in winter. They live for many years.

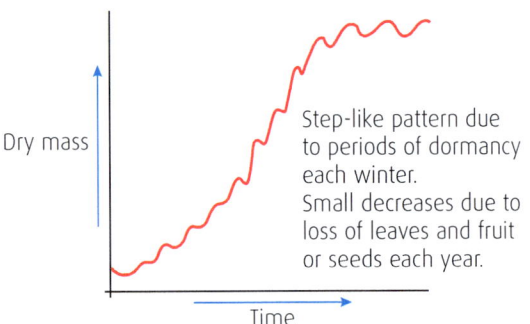

Deciduous tree growth curve

LIFE CYCLE

Plants show a number of stages in order to complete their life cycle. These are shown in the diagram.

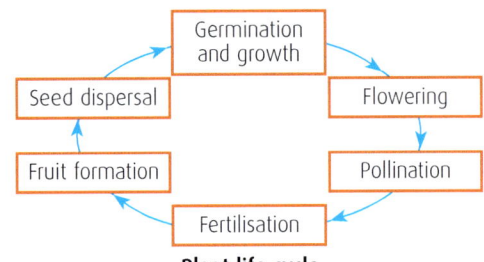

Plant life cycle

Growth and Development in Different Organisms – Growth Curves, Growth in Plants & Growth in Animals U2

GROWTH IN ANIMALS

Animal growth takes place throughout the body rather than in localised areas. Different areas grow at different rates at different times, producing changes to body proportions. The diagram shows the proportions of areas of the body at different ages.

DON'T FORGET

Plant growth takes place in localised areas. Animal growth takes place throughout the body.

6 months   1 year   3 years   7 years   13 years   26 years
Changes in body proportions

ANIMAL GROWTH CURVES

EXAMPLE 1 Humans

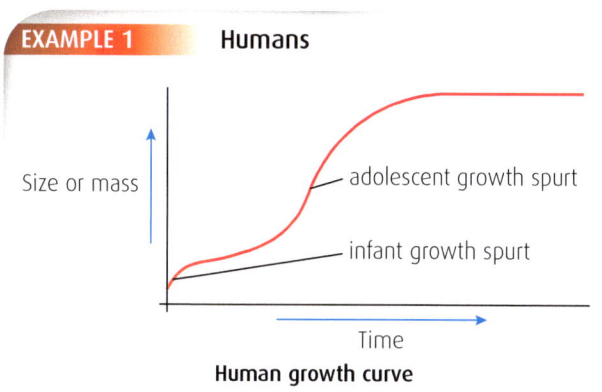

Human growth curve

EXAMPLE 2 Insects

Growth takes place in short spurts following the moulting of an old exoskeleton and before the new exoskeleton has hardened.

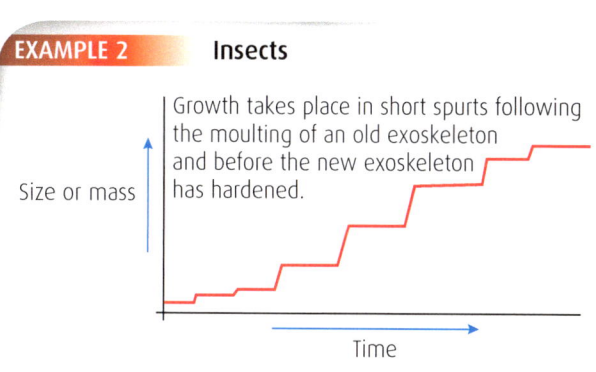

Insect growth curve

ANIMAL LIFE CYCLES

Some animals, such as insects, show very distinct stages in their life cycles which involve changes in appearance. This is called metamorphosis.

DON'T FORGET

Some other animals show metamorphisis. For example, frogs have a juvenile tadpole stage which looks very different from the adult.

EXAMPLE 1 Butterfly – complete metamorphosis

The juvenile stage looks different from the adult.

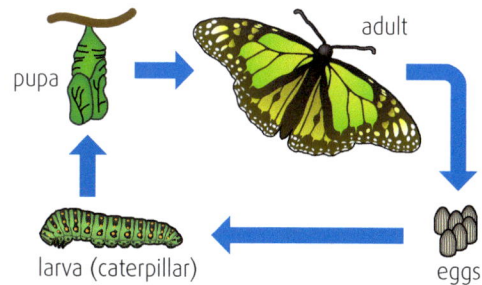

EXAMPLE 2 Head louse – Incomplete metamorphosis

The juvenile stage looks similar to the adult.

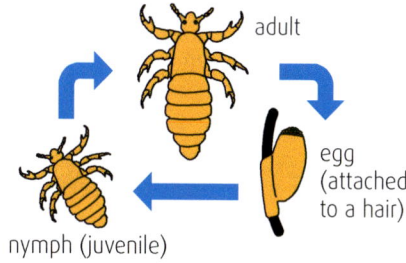

SUMMARY

- Growth is the result of an increase in cell number and an increase in cell size.
- Growth in plants is associated with specific areas. Growth in animals takes place throughout the body.
- The growth of most organisms follows a similar pattern. A period of slow initial growth is followed by a period of faster growth before growth slows and stops.
- The life cycles of both plants and animals show a number of stages.

THINGS TO DO AND THINK ABOUT

1. Name the two processes which contribute to the growth of an organism.
2. Describe how growth in plants differs from growth of animals.
3. What term is used for the change in appearance which takes place in some organisms during their life cycle?
4. Name the juvenile stage of a butterfly and of a head louse.

59

GROWTH AND DEVELOPMENT IN DIFFERENT ORGANISMS

FACTORS NEEDED FOR GROWTH IN PLANTS AND ANIMALS

FACTORS NEEDED FOR GROWTH IN PLANTS

CONDITIONS FOR SEED GERMINATION

Germination is the early stages in the growth of a seed. It can be used to investigate the conditions needed for growth.

The features of the investigation and the results are shown in the table.

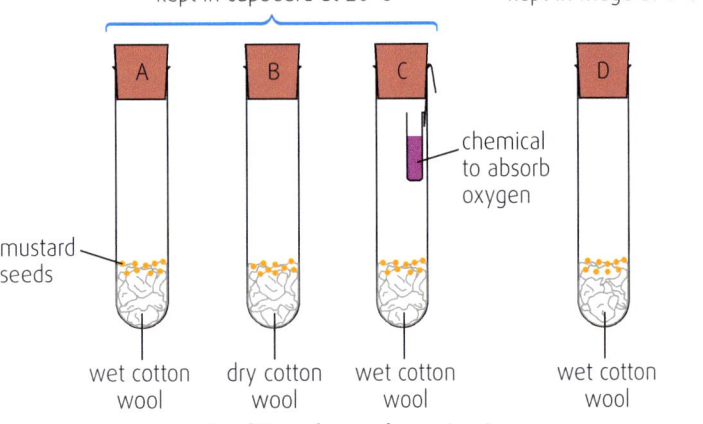

Tube	Conditions present	Missing factor	Successful germination
A	water oxygen warmth	none	yes
B	oxygen warmth	water	no
C	water warmth	oxygen	no
D	water oxygen	warmth	no

The results show that water, oxygen and warmth are all needed for successful germination and growth.

 DON'T FORGET

A suitable temperature, water and oxygen are all essential for seed germination.
Light is not necessary for germination but it is needed for continued plant growth.

OPTIMUM CONDITIONS FOR GROWTH

The optimum conditions for growth are those at which growth takes place at the maximum rate. Each factor that affects growth may have an optimum value. It can be found by comparing the growth rate at a range of values for that factor. All other conditions must be kept the same for a valid comparison to be made

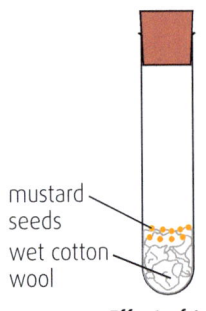

Five similar containers are set up. The containers are placed in different temperatures. After several days the numbers of seeds germinating are counted. Typical results for such an investigation are shown in the table and graph.

Effect of temperature on germination

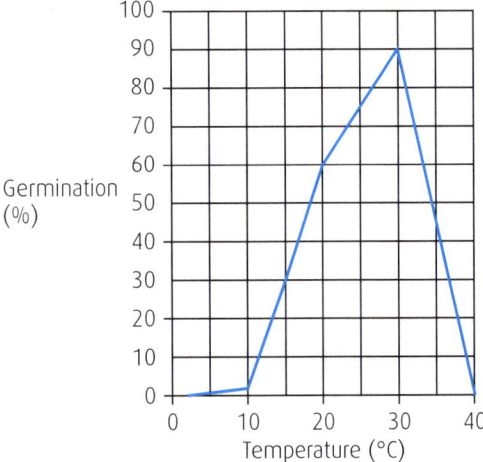

Temperature (°C)	Location	Germination (%)
1	fridge	0
10	chiller	2
20	room	60
30	incubator	90
40	incubator	0

Effect of temperature on germination

The results show that the optimum temperature for seed germination and growth is about 30°C.

Similar investigations can be carried out to find optimum values for other factors such as seed spacing and soil pH.

THE NEED FOR MINERAL ELEMENTS BY PLANTS

Some minerals are needed by plants in greater quantities than others. These are known as macro-elements. A deficiency in any of these elements causes reduced plant growth and other recognisable symptoms.

Growth and Development in Different Organisms – Factors Needed for Growth in Plants and Animals U2

Mineral element	Importance to plant	Deficiency symptoms
Nitrogen (N)	Part of many compounds including DNA and chlorophyll.	Yellowing of leaves (chlorosis)
Phosphorus (P)	Part of many compounds including DNA and chemicals involved in respiration and photosynthesis	Leaves darker than normal Red leaf bases
Potassium (K)	Needed for the activity of many enzymes	Yellowing of leaf edges
Magnesium (Mg)	Part of chlorophyll	Chlorosis

Importance of macro-elements to plants

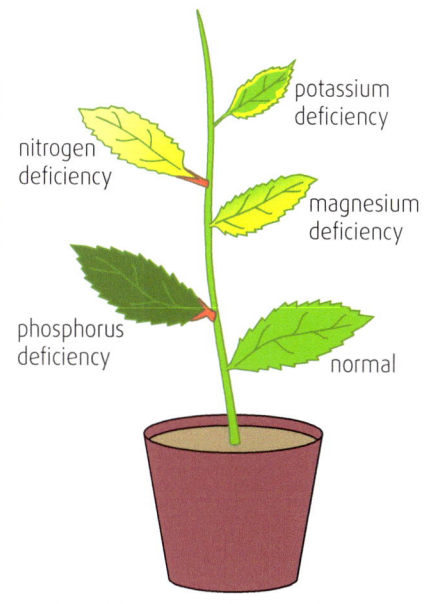

Effect of deficiency of macro-elements on appearance of plant leaves

COMMERCIAL PLANT GROWING

Commercial plant growers can alter the growing conditions of plants in order to increase their growth rates. Growing plants in enclosed conditions such as a glasshouse allows growers to control conditions such as light, temperature, carbon dioxide availability, water availability and nutrient availability (see page 50).

FACTORS NEEDED FOR GROWTH IN ANIMALS

A balanced diet is essential for the healthy growth of all animals, including humans. It must contain a number of major food components which have different functions. These are summarised in the table.

Component	Function	Sources
Carbohydrate	Supply energy	Root vegetables such as potatoes and carrots. Cereals such as wheat, rice and corn
Protein	Structure of cells. Enzymes to control cell reactions	Meat, fish, eggs, beans, nuts
Fat	Provide insulation. Structure of cells. Transport of some vitamins	Dairy products, meat, oily fish, vegetable oils

Major food components

In addition, small quantities of vitamins and minerals are needed. These have particular functions and are essential for good health. They are summarised in the table.

Vitamin	Function	Source
A	Maintains healthy eyes and prevents night blindness	Milk, eggs, liver, carrots
B1	Needed for normal nerve functioning and cell metabolism	Wholegrain cereal products, cauliflower, potatoes, oranges, liver and eggs
C	Needed for healthy skin and wound healing	Fruit, vegetables and liver
D	Needed for the absorption of minerals and the development of healthy bones	Dairy foods, eggs. It is also made in the skin during exposure to sunlight

Mineral	Function	Source
Calcium	Needed for the formation of bones and teeth	Dairy foods, green vegetables
Iron	Needed for the transport of oxygen in the body	Red meat, eggs, tuna fish, wholegrain cereals

SUMMARY

- A suitable temperature, water and oxygen are needed for the germination of seeds.
- The conditions for plant growth can be more easily controlled by growing them in glasshouses.
- The optimum value of a growth factor is the value at which growth is fastest.
- For healthy growth, a plant needs carbon dioxide from the air, water from the soil and a variety of minerals from the soil.
- For healthy growth, an animal needs a balanced diet containing correct quantities of carbohydrates, proteins and fats as well as a range of vitamins and minerals.

THINGS TO DO AND THINK ABOUT

1. Give four growth conditions of plants which can be controlled by growing them in glasshouses.
2. Explain what is meant by the phrase 'the optimum temperature for seed germination'.
3. What is 'chlorosis'? Name two minerals which are needed to prevent the condition.
4. Name one vitamin and one mineral which are both needed for the growth of healthy bones.

GROWTH AND DEVELOPMENT IN DIFFERENT ORGANISMS

THE EFFECT OF RADIATION AND CHEMICALS ON GROWTH

Some environmental factors can cause problems in growth and development.

Mutations are mistakes which happen during cell division. They cause changes to the genetic information of the daughter cells and are normally harmful. Mutations happen naturally but some factors can increase the chance of them happening. These factors are called mutagens. They include some chemicals and some forms of radiation.

RADIATION

There are many forms of radiation which are important and useful such as light, radio waves, microwaves and heat. Some forms of radiation, although useful, are also damaging to living tissue because they can cause mutations.

Examples of mutagenic radiation include ultraviolet light (UV), X-rays and gamma rays.

Mutations are particularly dangerous if they happen during the formation of sex cells because the mistakes can be inherited by any offspring.

People who work with radiation, for example staff who operate X-ray equipment, wear aprons with a lead lining to protect themselves from too much exposure.

Damage to cells caused by radiation can also lead to uncontrolled cell division and the formation of cancerous tumours. The tumours invade neighbouring parts of the body and can spread to other parts through the lymphatic system.

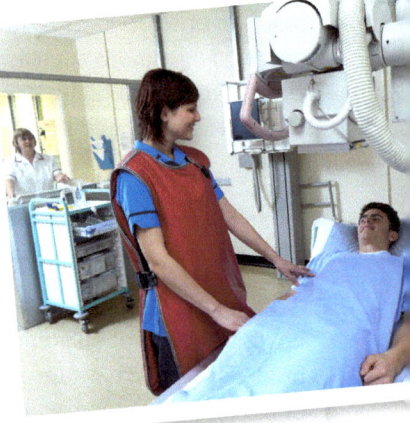

Protective X-ray apron

CHEMICALS

Some of the chemicals which cause the most damage to growth and development are those which result from lifestyle choices such as smoking, excessive consumption of alcohol and poor diet.

SMOKING

Cigarette smoke contains many harmful chemicals including tar, carbon monoxide, nicotine and arsenic. The health problems they cause can be made worse because of the addictive nature of smoking. These include:

- heart disease from blockage of the coronary artery which carries blood to the heart muscle
- stroke from blockage of arteries which carry blood to the brain
- high blood pressure
- damage to the cleaning mechanism of the lungs, increasing the chance of infections
- decreased breathing efficiency
- decreased ability of the blood to carry oxygen
- cancer, including cancer of the lung, mouth, tongue, stomach and kidney
- miscarriage
- reduced birth weight of babies
- impaired mental and physical development of babies from mothers who smoke when pregnant.

Effect of smoking on lungs

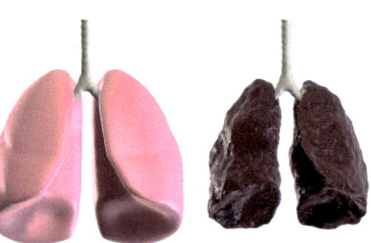

Lung from a non-smoker   Lung from a smoker

DON'T FORGET

Many of these effects can result from passive smoking, that is, breathing in smoke from a smoker close to you.

ALCOHOL

Alcoholic drinks vary greatly in their alcohol content. This makes it difficult to know exactly how much alcohol a person might consume. To make it easier, alcohol consumption is measured in 'units'.

Growth and Development in Different Organisms – The Effect of Radiation and Chemicals on Growth U2

Suggested maximum consumption is
- Men and women – 14 units per week; 3 units per day; at least 3 alcohol-free days per week
  The difference in amount is because of the difference in average body size.
- Under 18s and pregnant women – no alcohol.

> **DON'T FORGET**
> Alcohol is a high-energy substance and it can contribute to weight gain.

Drink	Beer	Cider	Alcopops	Wine	Fortified wine	Spirits
Alcohol content	4·4%	5·2%	5%	12%	15%	40%
Volume	1 pint	1 pint	275 ml	175 ml	175 ml	25 ml
Units	2·5	3	1·4	2	2·5	1

Alcohol content of a variety of drinks

The health effects of alcohol can be made worse by the addictive nature of alcohol. They include;
- respiratory distress – difficulty in breathing
- heart problems – disruption to the natural rhythm of the heart
- liver damage – some of this may be reversible but prolonged heavy consumption of alcohol can lead to permanent damage which will result in death
- impaired mental and physical development of babies from mothers who drink when pregnant.

Stages in liver damage

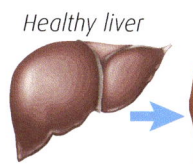

Healthy liver | Fatty liver | Liver fibrosis | Cirrhotic liver

Fat deposits lead to liver enlargement. Abstinence can lead to a full recovery | Recovery is possible but scar tissue remains | No recovery is possible

DIET

A balanced diet is a healthy diet. However, some components of our food, even though they are essential for health, can cause problems if taken in excess. They include fat, salt and sugar. Some processed foods can contain high levels of these substances.

- Fat is a high-energy food which can contribute to weight gain and obesity. Saturated fats, found in animal fats and dairy products, can cause circulatory problems leading to heart attacks and strokes. Unsaturated fats, found in some vegetable oils and oily fish, are more beneficial.

- Salt is needed to maintain the water content of blood and for the transmission of nerve impulses. It is used as a food preservative and for flavouring. High salt intake can lead to high blood pressure, increasing the risk of strokes and heart attacks.

- Glucose sugar is the end point of carbohydrate digestion. It is the main fuel used to provide energy in the body. A variety of sugars are added to many foods for flavouring. Their high-energy value can contribute to obesity, increasing the risks of heart attacks, strokes and diabetes.

SUMMARY
- Normal growth and development can be affected by a variety of environmental factors.
- Radiation leads to mutations by causing changes to DNA during cell division. These can be inherited if the change occurs during the formation of the sex cells. Radiation can also damage cells causing cancer.
- Tobacco smoke can cause a range of health issues including heart, respiratory and circulatory disorders and cancer.
- Alcohol can cause heart and respiratory disorders and liver damage.
- Too much fat, salt and sugar in our diet can cause health problems.

THINGS TO DO AND THINK ABOUT

1. Explain what is meant by the term 'mutation'.
2. Why is it especially important that the protective lead aprons worn by X-ray staff cover their reproductive organs?
3. How many pints of beer would lead to an alcohol intake just above the recommended daily maximum for a man?
4. Name two health problems which can be caused by a diet containing too much:
   (a) fat   (b) salt   (c) sugar.

MAINTAINING STABLE BODY CONDITIONS

THE NERVOUS SYSTEM

STRUCTURE OF THE NERVOUS SYSTEM

The nervous system is made up of three parts:

- the brain } central nervous system
- the spinal cord
- the nerves

THE CENTRAL NERVOUS SYSTEM

The brain and the spinal cord together make up the central nervous system (CNS). The CNS receives information from all the sensory receptors of the body. It processes this information and produces appropriate responses. Signals are sent from the CNS to relevant parts of the body (effectors) which carry out the responses.

Nerves carry signals from the sensory receptors to the CNS. They also carry signals from the CNS to the effectors (muscles and glands).

The brain is capable of conscious control of the body and of complex processes such as thinking, memory and decision making.

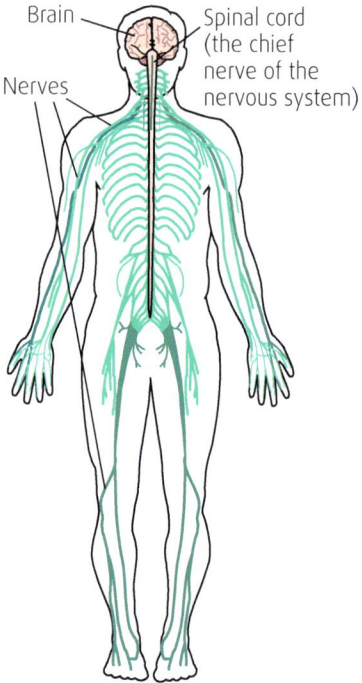

The nervous system

REFLEX ARC

A reflex action is an automatic response to a stimulus which happens without any conscious thought needed. A reflex arc illustrates the way in which the CNS and the nerves are connected to allow a reflex action to take place.

The stages of a reflex arc are:

- A stimulus is detected by a receptor. In this case heat is detected by receptors in the skin.
- A single sensory nerve cell carries an electrical signal from each receptor into the CNS.
- A signal passes to a motor nerve cell and is carried by it to an effector such as a muscle or a gland.
- In this case the effector is a muscle which responds by contracting to move the hand away from the heat.

Other nerve cells in the CNS can carry signals from the reflex arc to the brain so that it is aware of the event even though it is not involved in making the response.

Many reflex actions are protective, such as the withdrawal reflex illustrated above. Others allow body functions to adapt to circumstances, such as increasing or decreasing the heart rate to match body activity.

EXAMPLES OF REFLEX ACTIONS

The illustrations below show two reflex actions which you may be able to observe easily.

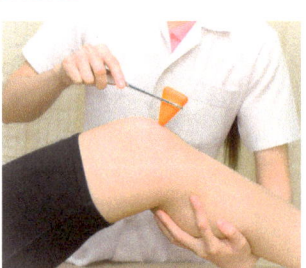

Knee jerk reflex

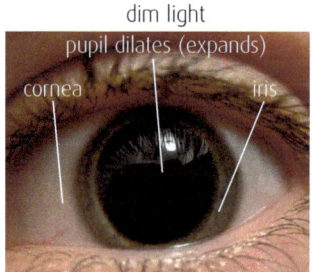

Pupil reflex

Maintaining Stable Body Conditions – The Nervous System | U2

OUR SENSES

Information enters the nervous system from sensory receptors which are found in all parts of the body. There are several types of sensory receptors which together make up our senses. Each type is able to detect a specific type of stimulus and produce an electrical signal which is then carried by the nerves. Some sensory receptors are found together in specialised sense organs whereas others are found less localised.

DON'T FORGET

The brain is the only part of the nervous system capable of conscious control of the body. The spinal cord can initiate responses without involving the brain, but they are purely automatic.

Other receptors detect various conditions within the body such as the carbon dioxide concentration of the blood, blood temperature, presence of food in the digestive system and the fullness of the bladder as it stores urine. Many of these receptors are involved in processes which maintain stable body conditions.

Sense	Stimulus detected	Where found
Sight	Light	Eyes
Hearing	Sound	Ears
Touch	Pressure	Skin and mouth
Taste	Dissolved chemicals	Tongue
Smell	Gaseous chemicals	Nose
Balance and movement	Movement of hairs caused by movement and changes of position of the head	Inner ears
Temperature	Heat movement to or from skin	Skin
Pain	Damage to nerves or body tissues	Skin, joints, bones and body organs
Position	Relative position of parts of the body	Throughout the body

The table gives information about some of our senses.

SENSE OF TOUCH

Our sense of touch is very sensitive. It is capable of discriminating between all sorts of different textures and pressures. However, it is not uniform over the whole body.

ACTIVITY

Working with a partner and using a divider, or even a paper clip bent into a U-shape, it is possible to compare the sense of touch on different parts of the body.

- Move the points of the divider, or ends of the paper clip, so that they are 1 mm apart.
- With your partner's eyes closed, gently touch the selected area of skin with the divider so that both points touch at the same time.
- Ask them if they feel 1 or 2 points of contact. They should feel only 1.
- Move the points of the divider apart by another 1 mm and try again.
- Repeat until your partner reports 2 points of contact and note the distance apart of the points of the divider.
- Repeat the whole procedure on a different area of skin.
- Make a table of your results.

Area of skin	Distance of 2 point detection (mm)
Finger tip	
Cheek	
Palm of hand	
Forearm	
Upper arm	
Calf	

SUMMARY

- The brain and spinal cord together make up the Central Nervous System which processes information from the sensory receptors and initiates responses to be carried out by effectors.
- Sensory nerves carry signals from the receptors to the CNS.
- Motor nerves carry signals from the CNS to the effectors.
- Reflex actions are automatic responses to a stimulus. They are rapid and help to protect the body or to control regular essential processes.

THINGS TO DO AND THINK ABOUT

1. What is the general name given to:
   (a) structures which detect stimuli and produce electrical nerve impulses;
   (b) structures which respond to nerve impulses from the CNS?
2. What are the two general functions of reflex actions?
3. Which two of our main senses detect chemicals?
4. Name the nerve cells which:
   (a) carry nerve impulses into the CNS;
   (b) carry nerve impulses away from the CNS.

MAINTAINING STABLE BODY CONDITIONS

HOMEOSTASIS & MAINTAINING BODY TEMPERATURE AND BLOOD GLUCOSE LEVELS

HOMEOSTASIS

Homeostasis means keeping the internal conditions of the body at suitable stable levels. It involves constant monitoring of the conditions and taking appropriate actions to reverse any change away from the ideal level.

NEGATIVE FEEDBACK

Negative feedback is a mechanism involved in controlling many aspects of homeostasis. It works like a thermostat controlling the temperature in a room. If the room temperature falls below a minimum point, the thermostat detects this and responds by switching on a heater. When the temperature increases past a maximum point, the thermostat switches off the heater. The system is automatic and keeps the temperature close to the ideal.

> **DON'T FORGET**
>
> Negative feedback mechanisms work by detecting a change and then responding by causing an opposite change.

Thermostat negative feedback

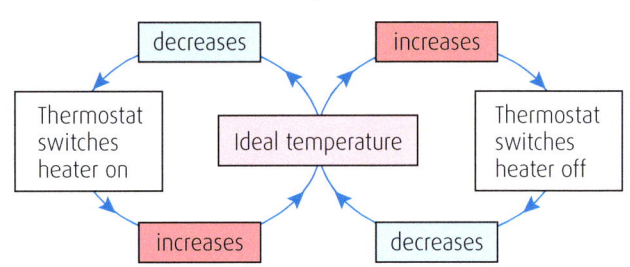

MAINTAINING BODY TEMPERATURE

The ideal body temperature for a human is about 37°C. If it rises above 40°C it causes heatstroke. If it falls below 35°C it causes hypothermia. Both can result in coma and death. The thermostat which controls body temperature is in the brain where it monitors the temperature of the blood flowing through it. If the temperature varies away from the ideal, nerve signals are sent to various parts of the body. Responses by the body counteract the change in temperature.

Control of body temperature

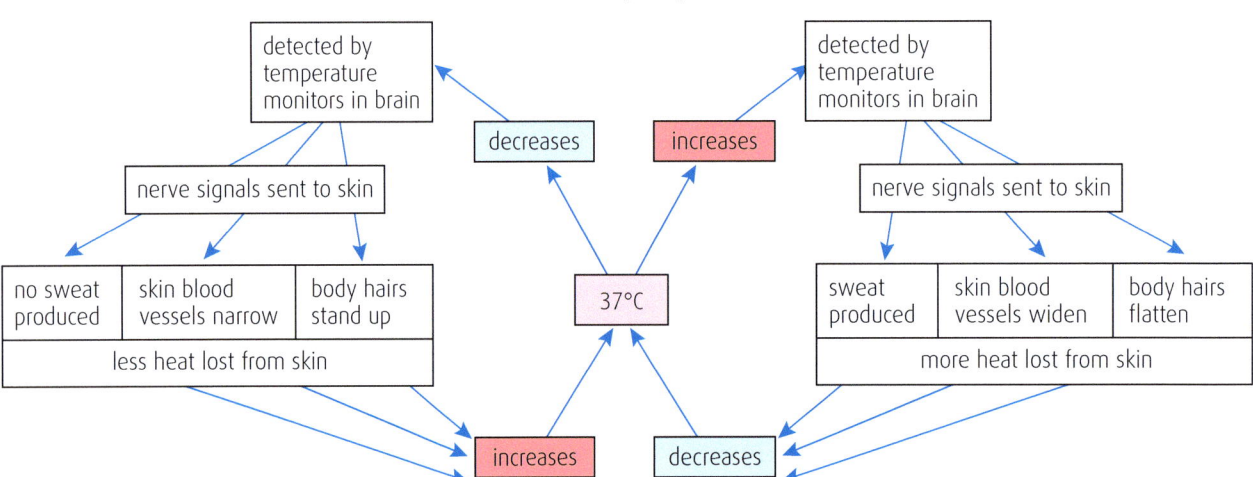

There are also other actions, not associated with the skin, which are involved.

If the body temperature decreases, signals are sent to various muscles causing them to contract repeatedly, making the body shiver. The metabolic rate will increase and more heat is generated in the body, replacing heat that has been lost.

If the body temperature increases, shivering stops. The metabolic rate will decrease causing less heat to be generated in the body, allowing it to cool down.

MAINTAINING BLOOD GLUCOSE LEVELS

Blood sugar (glucose) is produced by the digestion of carbohydrate foods and is absorbed into the blood from the intestines. Sometimes there will be a lot of glucose entering the blood and at other times there will be none at all. The concentration of glucose in the blood should be kept within narrow limits as too much or too little can cause severe medical problems.

The negative feedback mechanism involves the nervous system and hormones produced by glands in the pancreas.

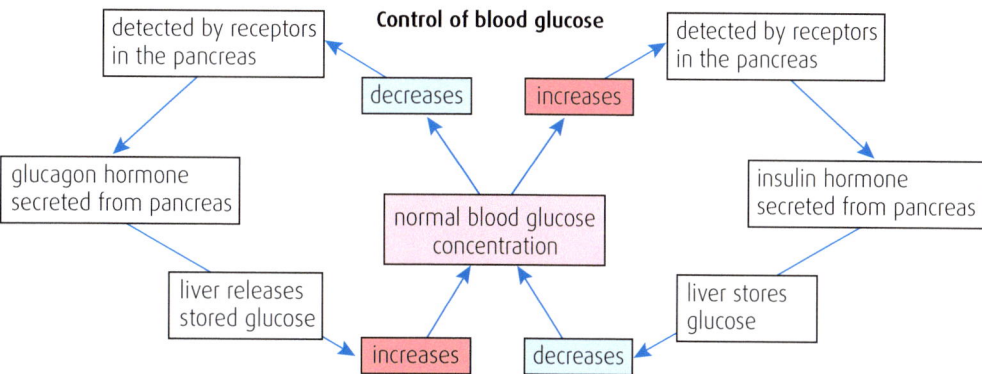

DIABETES

Diabetes is a medical condition in which the body cannot control blood glucose levels. There are two main forms of diabetes, Type 1 and Type 2.

- Type 1 diabetes is the more serious form and is caused by the pancreas not being able to produce any insulin. It is treated by regular injections of insulin. This causes the liver to store blood glucose and prevents the levels becoming too high. Care must be taken to eat sensibly so that the insulin injections do not cause the blood glucose levels to fall too far.

- Type 2 diabetes is caused by the pancreas not producing enough insulin. Type 2 diabetes develops gradually with age. Lifestyle factors such as obesity can increase the chance of developing this form of diabetes. It can often be treated by diet and exercise, but sometimes medication and insulin injections may be necessary.

HORMONES

Hormones are chemicals produced by a number of different glands in the body. A hormone enters the blood and is carried throughout the body. It causes a response at a different part of the body. Hormones provide longer-lasting responses than those produced by the nervous system. They are important for controlling many aspects of homeostasis and for controlling the development of the body.

DON'T FORGET

A hormone produces a longer lasting effect than a nerve signal.

SUMMARY

- Homeostasis means keeping the internal conditions of the body stable.
- Many of the conditions are maintained by negative feedback mechanisms. These work by making a response to a change in a condition. The response causes an opposite change which returns the condition to normal.
- Homeostatic actions involve the nervous system and may also involve hormones.
- Diabetes is a medical condition in which the pancreas produces no insulin or too little insulin. This affects the control of blood glucose levels and can be dangerous if not treated.

THINGS TO DO AND THINK ABOUT

1. What is meant by the term 'homeostasis'?
2. Explain why the term 'negative feedback' is a suitable description for the mechanism.
3. The body responds to a decrease in body temperature in a number of ways. Give one way it responds by reducing heat loss and one way it responds by increasing heat production.
4. Where is the hormone insulin produced and where does it show its effect?

UNIT 3: LIFE ON EARTH

ANIMAL AND PLANT SPECIES DEPEND ON EACH OTHER

BIOMES, ECOSYSTEMS & BIOTIC FACTORS AFFECTING ORGANISMS

Biomes and ecosystems both refer to ecological systems.

BIOMES

Biomes are the large ecological regions of the Earth. Each biome has its own characteristic climate and community of plants and animals adapted to survive there.

The main terrestrial biomes are shown on the map.

Some of these are described below.

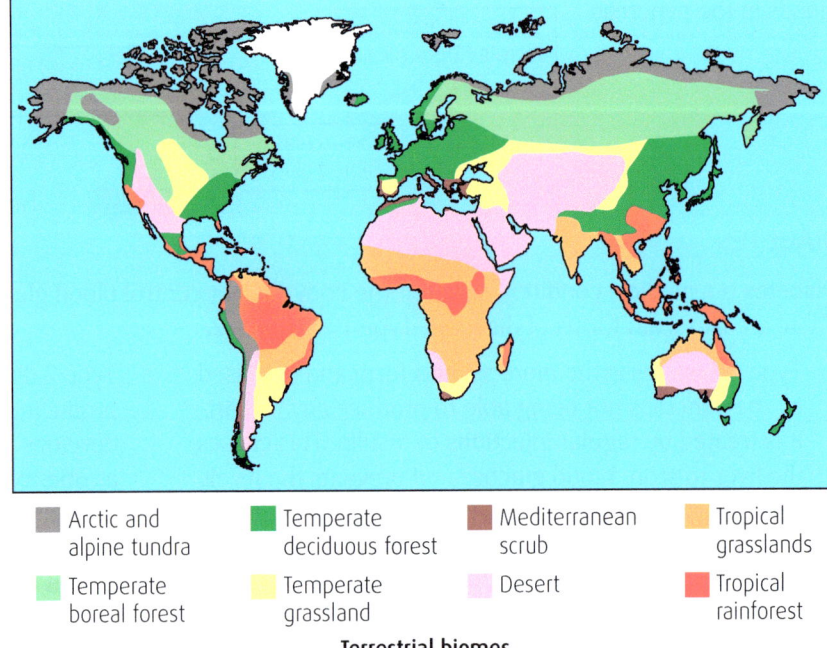

Terrestrial biomes

TUNDRA

Tundra is the most northerly and the coldest region of vegetation. It has frozen soil, low rain or snow fall, a short growing season and low levels of plant nutrients. The vegetation consists of lichens, mosses and other plants which can survive the extreme conditions.

TEMPERATE BOREAL FOREST

This is the great area of coniferous forests (cone-bearing trees) of Canada and northern Europe. It has long, cold, dry winters and short, warm, moist summers. It forms the largest terrestrial biome.

TEMPERATE DECIDUOUS FOREST

This is the broad-leaved forests south of the coniferous forests. It has a more moderate climate with a longer growing season. Much of this biome has disappeared because of human settlement.

TEMPERATE GRASSLAND

This biome includes the prairies of North America, the pampas of South America, the steppes of Europe and Asia and the veld of South Africa. It has hot summers, cold winters and moderate rainfall. Much of this biome is used for agriculture.

TROPICAL RAINFOREST

Tropical rainforests are found in hot wet regions of Central and South America, Africa and the islands of south-east Asia. They contain the most varied communities of plants and animals. Seasons vary from wet to dry with little other change.

DESERT

Deserts are extremely dry and can have wide variations of temperature. They support a very limited range of highly adapted plants and animals.

ECOSYSTEMS

Ecosystems are smaller, more localised systems of a habitat and its community of organisms living there. The organisms are all part of a single food web. A full description of an ecosystem includes the various abiotic (non-living) factors which characterise it.

DON'T FORGET

Biomes are large-scale geographic areas determined by climate and main type of vegetation. Ecosystems are smaller units of a habitat together with its community of plants and animals.

Animal and Plant Species Depend on Each Other – Biomes, Ecosystems & Biotic Factors Affecting Organisms U3

BIOTIC FACTORS AFFECTING ORGANISMS

Biotic factors are the effects of living organisms on other organisms. Examples are feeding relationships, competition, disease and by providing a habitat. For example, a tree provides a habitat for many smaller organisms. These must be distinguished from abiotic factors which are the physical conditions such as temperature, humidity, light and oxygen concentration which affect organisms.

PREDATION

Predation is when one organism, the predator, kills and eats another organism, the prey.

In a stable ecosystem the numbers of both predator and prey species fluctuate in such a way that both species survive. The predators are dependent on the prey as a source of food, but the prey species is also dependent on the predators. Without the predators, the number of prey would increase to a point where they would run out of food.

The diagram shows how the populations of predators and prey can fluctuate.

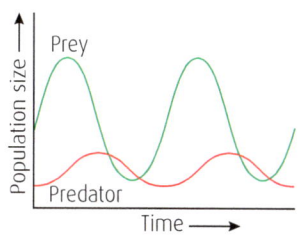

A high prey population means lots of food for the predators so their numbers increase. As this happens, the number of prey decreases.
This means less food for the predators causing their numbers to decrease.
The number of prey will then be able to increase again.

COMPETITION

Competition occurs when organisms require the same, or similar, resources for their survival. These resources include food, water, territory, light and reproductive partners. Some of the organisms involved will be better able to gain the resource and will have a better chance of survival than those that are less-able.

- **Interspecific competition** – Competition can occur between different species. When this happens, one species is usually a better competitor for the shared resource than the other. This will lead to a decrease in the population of the less-able species.
  For example, red and grey squirrels can both live in mixed woodland containing broad-leaved trees and coniferous trees. In this case, the grey squirrel is the better competitor and the red squirrels will eventually disappear from the habitat. In coniferous woodland it is the red squirrel which is the better competitor and so it can survive in areas where grey squirrels can't.
- **Intraspecific competition** – This is competition between individuals of the same species. Competitive success depends on variations between individuals. It ensures that those individuals best adapted to the environment have the best chance of surviving and reproducing. For example, robins are very territorial and a male robin will defend his territory against other robins. This ensures that he controls an area with sufficient food for himself and his offspring. His red breast is a warning to other robins. Singing at points around his territory is usually enough to keep other robins away, but he will attack them if his warnings are ignored.

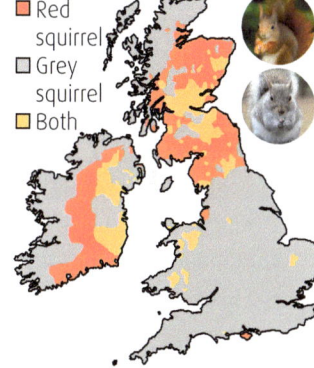

Distribution of red and grey squirrels

Robins sing to defend their territory

SUMMARY
- Biomes are large geographic regions, each with its own characteristic climate and vegetation.
- Ecosystems are smaller systems consisting of a habitat and the community of living organisms which occupy it.
- Biotic factors are the effects that living organisms have on other organisms. They include feeding relationships, competition and diseases.

THINGS TO DO AND THINK ABOUT

1. A freshwater pond is an ecological system. Would it be described as a biome or as an ecosystem? Give a reason for your answer.
2. What is the difference in the main type of vegetation of temperate boreal forest and temperate deciduous forest?
3. Foxes eat rabbits. Explain why a population of rabbits depends on foxes for their survival.
4. Explain why an infectious disease should be considered to be a biotic factor.

ANIMAL AND PLANT SPECIES DEPEND ON EACH OTHER

SAMPLING TECHNIQUES & THE EFFECT OF ADDING OR REMOVING SPECIES ON OTHER SPECIES

SAMPLING TECHNIQUES

The size of a population of organisms in an ecosystem can be estimated by taking samples of the population. It is important that the samples are representative of the area as a whole and that enough samples are taken to make the results reliable.

Different sampling techniques are available to suit different types of organisms.

QUADRATS

A quadrat is a square frame, usually measuring 0·5 m × 0·5 m, which equals 0·25 m². Quadrats are used to sample stationary organisms such as daisies in a lawn or limpets on rocks.

To estimate the number of daisies in a lawn, the following method is used.

1. Measure the area of the lawn in m².
2. Use quadrats at random at several positions on the lawn.
3. Count the number of daisies in each quadrat and calculate an average.
4. Calculate the estimated total number by multiplying the average by 4 (to give the number per m²) and then by the area of the lawn.

TRANSECTS

Transects involve placing quadrats at regular intervals along a line. They can be used to investigate the effect of a change in an abiotic factor on the distribution of suitable organisms in an ecosystem.

For example, a transect may run from a shaded area into open ground to investigate the effect of increasing light intensity on the distribution of plants. Another example would be to investigate the different organisms present between the high and low tide marks on a shore.

PITFALL TRAPS

These are used to sample small invertebrates.

Some precautions must be taken if they are to give useable results:

- The rim must be level with the soil surface to allow animals to fall in.
- The trap must be covered to stop rain getting in and to prevent birds eating trapped animals.
- There must be a gap between the cover and the soil to allow animals to reach the trap.
- The trap should have small holes in the base to allow any rain water to drain away.
- The trap must be checked frequently to collect results before some of the trapped animals eat others.
- There should be enough traps set to obtain representative and reliable results.

A quadrat

The diagram represents a lawn and the position of 10 quadrats. The number of daisies present in each quadrat is shown.

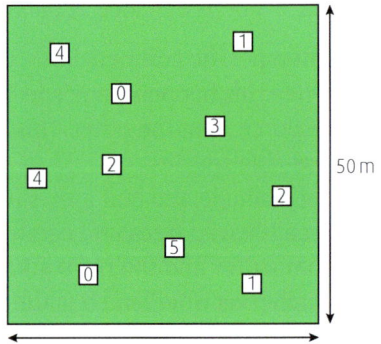

Area of lawn = 50 m × 50 m = 2500 m².
Total number of
daisies counted = 4+1+0+2+3+4+2+0+5+1
= 22
Average number of daisies
per quadrat = 22 ÷ 10 = 2·2
Estimated total number of
daisies on lawn = 2·2 × 4 × 2500
= 22000

A pitfall trap

DON'T FORGET

To produce useable results, samples must be representative of the whole area and be reliable. This is usually achieved by sampling at random and by taking several samples, the more the better.

Sampling Techniques & the Effect of Adding or Removing Species on Other Species U3

THE EFFECT OF THE ADDITION OR REMOVAL OF SPECIES ON OTHER SPECIES

In an ecosystem, all of the organisms which make up the community are part of food chains which begin with green plants and end with top predators. The food chains connect together to make a food web. In this way, all of the organisms are part of the same system.

If a new species was introduced into this system, or if a species was completely removed, then it would have knock-on effects on many of the other species present.

A well-established ecosystem with a complex food web is better able to adapt to such changes. Less-complex food webs with fewer interconnections are more likely to suffer severe disruptions.

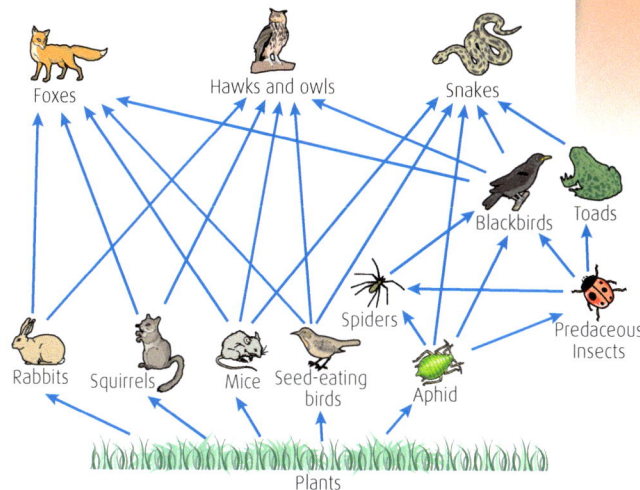

A food web

KILLER SHRIMPS

Killer shrimps are small freshwater shrimps which have been introduced into Britain from Eastern Europe on the equipment of water sport enthusiasts. They kill many native species of small fish and invertebrates and have been responsible for the extinction of some species from some areas.

A pack of wolves

WOLVES

Wolves are top predators. Their presence controls the numbers of other species in an area. The extinction of wolves from the Yellowstone National Park in the USA allowed the population of elks to increase. This caused severe overgrazing and the loss of some plant species. Other species which were dependent on these plants were badly affected.

Warning notice for killer shrimps

SUMMARY

- The number of a species in an ecosystem can be estimated by taking samples.
- Samples must be representative of the whole area and the results must be reliable. This is usually achieved by taking several random samples.
- All the species in a community of organisms are part of a single food web. The loss or addition of any one species can affect all the others.
- Large complex food webs are more stable and can adapt to change better than simpler food webs.

THINGS TO DO AND THINK ABOUT

1. Explain the benefits of sampling randomly and by taking as many samples as possible.
2. The diagram represents a lawn and the position of several 0·25 m² quadrats.
   The number of dandelions present in each quadrat is shown.

   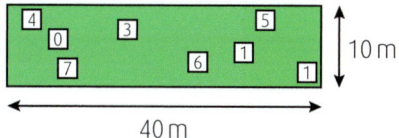

   Estimate the total number of dandelions in the lawn.
3. Describe a possible effect of the removal of rabbits on the population of owls in the food web shown above. Explain your answer.
4. Describe a possible effect of the removal of toads on the population of blackbirds in the food web shown above. Explain your answer.

71

IMPACT OF POPULATION GROWTH AND NATURAL HAZARDS ON BIODIVERSITY

IMPACT OF POPULATION GROWTH ON BIODIVERSITY 1

HUMAN POPULATION GROWTH

Humans have existed on Earth for about 50 000 years. For most of this time the total population has increased very slowly. This began to change in the 1800s when industrialisation and agricultural changes led to increases in wealth and food production.

It took over 49 000 years for the population to reach 1 billion in about 1800. It took only another 132 years for it to reach 2 billion in 1932. In 2011 the population reached 7 billion.

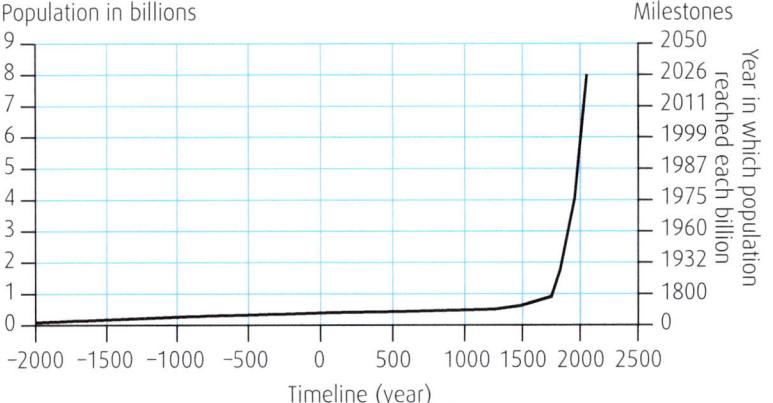

World population growth

EFFECT OF POPULATION GROWTH ON BIODIVERSITY

Biodiversity refers to the range of different species present in an ecosystem. The greater the biodiversity, the healthier and more stable is the ecosystem.

The growth of the human population has been possible because of increases in food production. An increase in the use of fertilisers and pesticides has contributed to this. Both of these can reduce biodiversity.

- Fertilisers can be washed from the soil into rivers and lochs increasing the growth of aquatic algae, producing an algal bloom. Bacteria feed on the dead algae and increase in number, using up most of the dissolved oxygen. This causes the death of fish and other aquatic organisms.
- Pesticides can kill organisms other than the pests they are meant for, reducing biodiversity.
  Some pesticides are persistent. This means that once they enter the tissues of an organism, they remain there. They can be passed on through the food chain and accumulate in the bodies of larger predators.
  One example is DDT. Its use as an agricultural insecticide is now banned throughout the world. However, it is still used in some countries to kill insect pests such as mosquitoes which carry the disease malaria.
  Despite being banned in the USA in 1972, DDT was found to be present in almost all human blood samples tested there in 2005. DDT was banned in Britain in 1984.

Algal bloom in Lake Erie in 2011

ECOLOGICAL FOOTPRINTS

All the resources used by humans come from planet Earth. These include food, energy, water and materials for manufacturing processes. The Earth must also absorb all the waste material which we produce. Together, these represent a demand which we put onto the Earth. The ecological footprint is a measure of this demand.

The Earth has an ability to provide the resources we need and to cope with our wastes. The measure of this ability is called its biocapacity.

By comparing the ecological footprint with the biocapacity, it is possible to calculate

Spraying crops with DDT in the USA before the ban

DON'T FORGET

Biodiversity refers to the range of species existing in an ecosystem

Impact of Population Growth on Biodiversity 1　U3

if the Earth is able to meet our demands. The results can be expressed as a figure representing how many planet Earths would be needed to meet our demands.

The graph shows that, at present, we would need about 1·4 planet Earths to supply all our needs. If human societies continue to develop as they have been doing, by the year 2050 it would take almost three planet Earths. The graph also shows how much the ecological footprint needs to decrease to reach a sustainable level.

This figure changes with time because the ecological footprint is affected by many factors including:

- increasing human population
- improving technology, making energy use more efficient
- recycling materials
- increasing wealth, creating more demand for consumer products.

Some of these factors help to reduce the ecological footprint by reducing our demand on the Earth's resources. Some of these factors increase the ecological footprint by increasing our demand.

At present the ecological footprint is greater than the Earth's biocapacity. This has a number of consequences:

- It reduces biodiversity because the more of the Earth's resources we use, the less are left for other organisms. This is mainly due to the destruction of habitats. An example of this is the clearing of rainforests to provide land for growing crops.
- It is unsustainable. Using more resources from the Earth than it is capable of producing is using up the natural reserves of these resources. Eventually this will result in disasters such as widespread starvation and failure of energy supplies.

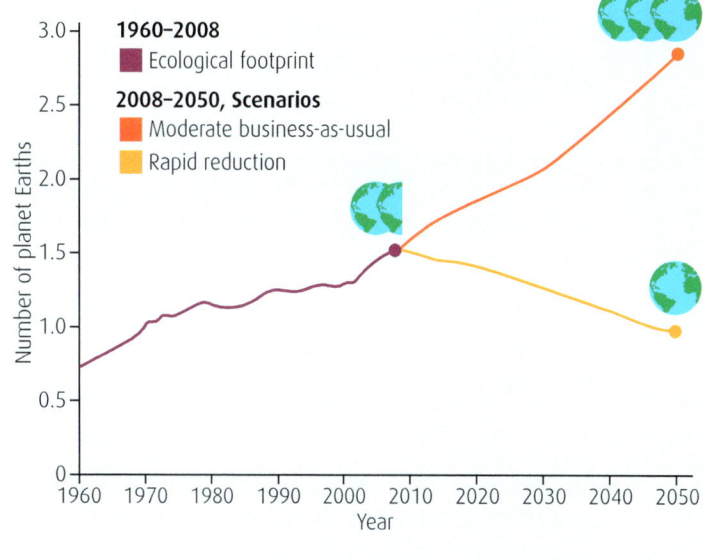

Number of planet Earths needed to meet our demands

SUMMARY

- The human population increased slowly until the 1800s. It then began to increase rapidly because of increased food production.
- Increasing land use for agriculture, increasing use of fertilisers and increasing use of pesticides all reduce biodiversity.
- At present the ecological footprint is greater than the Earth's biocapacity. This is not sustainable and the ecological footprint must be reduced to avoid future disasters.

 ### THINGS TO DO AND THINK ABOUT

1. Using the graph of world population growth, describe what has happened to the length of time taken to reach each additional 1 billion increase in population.
2. Explain why DDT is considered to be more dangerous when used as an agricultural insecticide rather than as a way of preventing the spread of disease.
3. Give five different demands that humans make on planet Earth and which contribute to the ecological footprint.
4. For each of the following factors, say whether they cause an increase or a decrease in the ecological footprint:
   (a) increasing human population
   (b) improving technology, making energy use more efficient
   (c) recycling materials
   (d) increasing wealth, creating more demand for consumer products
   (e) increasing use of renewable energy
   (f) increasing use of biological detergents, allowing clothes to be washed at lower temperatures.

 **DON'T FORGET**

The Ecological Footprint is a measure of all the demands humans make on the earth's resources. Biocapacity is a measure of the earth's ability to cope with these demands.

IMPACT OF POPULATION GROWTH ON BIODIVERSITY 2

HABITAT DESTRUCTION

Habitats change naturally because of changing environmental conditions. Such changes normally happen slowly and organisms may adapt or evolve with the changes.

Humans have always altered their environment and have affected biodiversity. The rapid increases in the human population mean that changes are now happening on a greater scale that ever before.

DEFORESTATION

Deforestation is the removal of trees from an area faster than natural regeneration can replace them. This happens for a number of reasons including:

- Groups of people burn areas of forest to clear space for growing crops. The soil in such areas is often poor and soon becomes unsuitable for crop growth. The people then move on and clear another area of forest.
- Commercial companies are responsible for clearing large areas of natural forest for timber production or mining. This is sometimes done illegally.

Deforestation

INTENSIVE AGRICULTURE

Intensive agriculture involves the clearing of land to allow large scale cultivation of crops. It destroys the natural habitat and reduces biodiversity. Intensive agriculture also involves the use of fertilisers and pesticides which can have other adverse effects (see pages 56 and 72).

ACID RAIN

All rain is slightly acidic because it contains dissolved carbon dioxide from the air. Acid rain is more acidic because it contains pollutants such as dissolved sulphur dioxide and oxides of nitrogen. Acid rain can make soil too acidic for the growth of many plants and can cause physical damage to plants. It can affect waterways, killing aquatic animals and plants.

Sign in Canada warning of the effects of acid rain

OVER-FISHING

Over-fishing is the catching of fish on a scale which reduces the population to a level where it cannot be maintained. This puts them in danger of extinction. Other species which are caught by accident may also be affected. Over-fishing disrupts the natural food chains and reduces biodiversity.

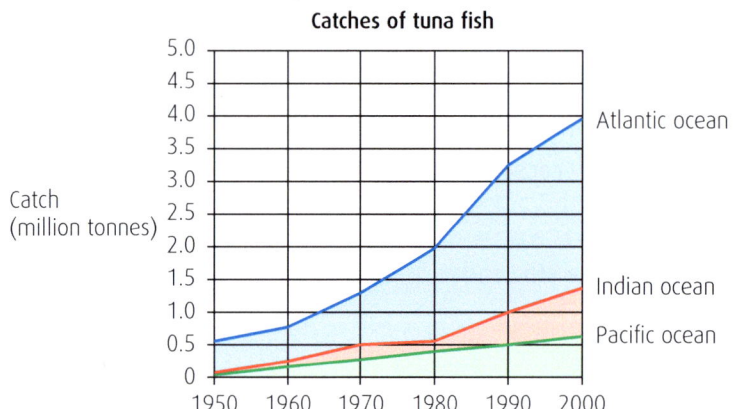

Cartoon highlighting the dangers of overfishing

The graph shows the increase in tuna caught from three different oceans between 1950 and 2000.

Impact of Population Growth on Biodiversity 2 — U3

OTHER EFFECTS OF HUMAN ACTIVITY ON BIODIVERSITY

CLIMATE CHANGE

Climate change has always taken place. However, in recent times human activity is causing global warming which is resulting in rapid climate change. The burning of fossil fuels is releasing carbon dioxide into the atmosphere in such quantities that its concentration is increasing. Carbon dioxide acts as a greenhouse gas. This means that it reduces the amount of heat being lost from the Earth into space. The resulting rise in the temperature of the Earth's atmosphere and oceans is causing melting of polar ice. This leads to rising sea levels and loss of coastal habitats, so reducing biodiversity.

Rising sea temperatures also affects the distribution of fish species. Some North Sea species are becoming scarcer in this area because they are moving further north to cooler water. At the same time, species normally found further south in warmer water are moving into the North Sea.

OIL SPILLS

Oil spills happen from oil platforms and from tankers hitting rocks at sea. Oil destroys the waterproof properties of the fur of mammals and the feathers of birds. This means that they are unable to float and are likely to drown. It also reduces the insulating ability of the fur or feathers, causing hypothermia. If the animals try to clean themselves they will ingest the toxic oil and be poisoned.

SEWAGE

Sewage is a major cause of water pollution. Some coastal countries dump sewage into the sea and boats sometimes release untreated sewage into the sea. Rivers may be used as a way of getting rid of sewage.

The contamination of water with untreated sewage causes an increase in the number of bacteria which use the sewage material as food. The bacteria use up oxygen from the water causing the death of fish and other aquatic organisms, decreasing biodiversity.

The contamination of water with untreated sewage can spread diseases such as cholera and typhoid. Over three million people, mainly children, die every year from diseases caused by contaminated water.

DOMESTIC WASTE

Domestic waste contains a wide variety of materials including food, paper, textiles, electrical equipment, packaging and plastic bags. Increased recycling means that some of this material is reused, but much still ends up in landfill sites. Most of the waste will eventually be broken down by the action of bacteria, but it is a slow process with some materials taking many decades to decompose.

Decomposition in landfill sites produces methane. This escapes into the atmosphere or is collected and burnt, producing carbon dioxide. In either case a greenhouse gas is released, adding to climate change.

GENETIC POLLUTION

There is an increasing use of genetically modified crop plants. There have been cases of cross-pollination between such crops and neighbouring non-modified plants. Once this happens it cannot be reversed and so natural varieties of non-modified crop species may be lost.

Loss of biodiversity can cause many problems including:
- reduced chance of discovering new medicines
- reduction in biological materials such as fibres, dyes and oils that are used in industry
- reduction in benefits from leisure activities such as wildlife watching and hiking.

SUMMARY

Human activity can reduce biodiversity in many ways. These include:
- clearing land for agriculture or obtaining timber
- releasing pollutants which harm organisms or damage habitats
- overuse of fertilisers and pesticides which harm organisms or damage habitats
- reducing wild populations of animals to levels which endanger their survival
- contamination of non-modified plant varieties by genetically modified varieties
- disposal of sewage and domestic waste.

THINGS TO DO AND THINK ABOUT

1. Give two reasons for deforestation.
2. (a) What is meant by the term 'greenhouse gas'?
   (b) Give two examples of greenhouse gases.
3. Give two problems caused by the contamination of water with untreated sewage.
4. What causes the breakdown of materials in a landfill site?
5. How many times greater was the tuna catch from the Atlantic Ocean in 2000 than in 1950?

IMPACT OF POPULATION GROWTH AND NATURAL HAZARDS ON BIODIVERSITY

IMPACT OF NATURAL HAZARDS ON BIODIVERSITY

FOREST FIRES

Fire is important in maintaining the natural balance of many forest ecosystems. Forests tend to recover quickly from the effects of small-scale fires and the heat of fires is essential for the germination of some seeds. In recent years, climate change has meant that many forests are now much drier than they were and fire is now a major threat.

For example, tropical rainforests and forests which are often covered by clouds normally do not burn on a large scale because of the humid, moist conditions. However, in the 1980s and 1990s, these areas were devastated by wildfires. Large-scale fires make future fires more likely because they leave behind dead wood which increases the available fuel. They also allow colonisation by plants such as grasses which are more likely to catch fire.

The increase in forest fires may affect biodiversity in a number of ways:

- They are a significant source of carbon dioxide which contributes further to global warming.
- They reduce the mass of vegetation and can alter the water cycle. This can affect neighbouring marine ecosystems such as coral reefs.
- The smoke can significantly reduce photosynthesis and can be harmful to the health of human and animals.

New Mexico, USA forest fire in 2012

EARTHQUAKES

There are more than 3 million earthquakes every year. Most of these are too weak to notice. Stronger earthquakes can cause considerable damage to buildings in populated areas and to natural environments.

Almost 80% of all earthquakes happen along the rim of the Pacific Ocean. The same region also contains over 75% of the world's volcanoes.

Earthquakes can cause devastating side effects. On 26th December 2004, an undersea earthquake in the Indian Ocean produced a series of tsunamis which killed over 225 000 people in 14 countries. In January 2010 an earthquake in Haiti killed over 200 000 people.

Such powerful events can have significant effects on natural environments:

- The Sichuan region in China was hit by an earthquake in May 2008. More than 69 000 people were killed and 4.3 million people were left homeless. The area is a designated area for biodiversity conservation. It contains more than 12 000 plant species and 1122 species of vertebrates. It also includes over 50% of the habitat of the wild giant panda population and 60% of the panda population was affected. Over 23% of its habitat was destroyed and the remaining habitat was broken up into separate areas. This could reduce the chance of surviving pandas finding mates and reproducing.
- The coastline of Chile suffered from an earthquake and tsunami in 2010. Some beaches subsided and some rocky reefs were uplifted. Both of these effects reduced biodiversity. However, the earthquake also produced some benefits to the natural environment. Some of Chile's sandy beaches were being lost due to coastal erosion. The earthquake caused the uplifting of some coastlines, producing rocky seawalls which protect the beaches. The beaches are no longer being eroded and are being recolonized by species which had been lost.

Panda habitat destroyed by earthquake

Pandas rescued following earthquake

Impact of Natural Hazards on Biodiversity U3

TSUNAMIS

A tsunami is a series of extremely high waves that have been produced by the displacement of a large body of water. This can be caused by earthquakes, volcanic eruptions and landslides.

The tsunami of 26th December 2004 killed many people and devastated the lives of many more. It also caused large-scale environmental destruction. Coral reefs, coastal mangrove swamps and coastal forests were damaged and will take many years to recover.

Another important effect of a tsunami like this is the flooding of the land by sea water. The salt makes the soil unsuitable for crops and makes the recovery of the affected region more difficult.

26th December 2004 tsunami

WIND

The strong winds which occur during cyclones and tornadoes can cause extensive damage but this is usually short term and fairly localised. Ecosystems tend to recover quickly from this type of damage. Wind can have a more long-term effect when it causes soil erosion. Recovery is difficult because the remaining land is too infertile to allow recolonization by plants. This affects the food chains of the area.

Soil erosion is more likely in arid or semi-arid conditions when it can lead to desertification. Plants are important in reducing soil erosion because their roots hold the soil and prevent it being blown or washed away. The removal of natural plant cover through processes such as deforestation, overgrazing and the cultivation of unsuitable land leaves the soil at greater risk of erosion.

Ruined village following the tsunami

SUMMARY

Biodiversity can be adversely affected by natural events. These include:
- Forest fires – these are becoming more damaging because climate change is creating drier conditions in some forests.
- Earthquakes – these cause physical damage to habitats and can produce tsunamis which cause further devastation.
- Tsunamis – these cause devastation to coastal areas. They can make soil infertile for significant distances inland due to extensive flooding with salt water.
- Wind – strong winds can cause soil erosion in arid areas. This is more likely to happen following deforestation and the cultivation of unsuitable land.

Soil erosion due to intensive farming in arid regions

DON'T FORGET

Some of these natural hazards are inter-linked. For example, earthquakes can trigger volcanic eruptions and tsunamis.

THINGS TO DO AND THINK ABOUT

1. Why do some scientists think that forest fires are becoming an increasing threat?
2. Give two effects that the earthquake of 2008 had on the habitat of wild giant pandas.
3. What effect do tsunamis have on coastal land which makes it infertile for crop growth?
4. Give two human activities which can leave land more prone to soil erosion.

THE NITROGEN CYCLE

RECYCLING

THE ROLE OF MICROBES IN THE RECYCLING OF NUTRIENTS

Recycling of nutrients is a process which happens in nature. If it did not, the nutrients that plants and animals require for survival would soon be used up and life would cease to exist.

However, the chemical substances required by living organisms do not run out. This is because dead organisms, and the waste materials they produce during their lifetime, are broken down into simple substances which can be absorbed by future generations of living things.

A group of organisms known as decomposers are responsible for the breakdown of dead and waste material to release nutrients back into the ecosystem. Decomposers are mainly different types of fungi and bacteria and, although they are too small for us to see with the naked eye, they play an essential part in the recycling process.

The nutrients which are released by the action of fungi and bacteria are returned to the soil and are able to be absorbed by plants for growth.

Among other things, nitrogen is recycled in this way.

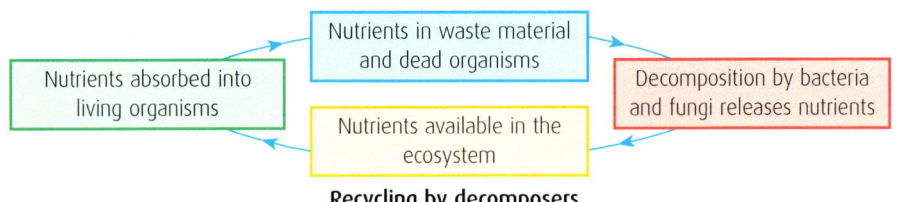

Recycling by decomposers

THE RECYCLING OF NITROGEN

All living organisms need nitrogen to make protein. Unfortunately, although the air around us is composed of almost 80% nitrogen, organisms are unable to use it in this form. Nitrogen is absorbed into plants in the form of chemicals called nitrates which are produced in the soil by the action of the decomposers.

When plants take up nitrates, they use the nitrogen they contain to make plant protein. This allows the plant to grow.

Animals cannot get their nitrogen from nitrates in the soil, so instead they eat the plants. Animals digest the plant protein into smaller molecules which are used to make animal protein.

Nitrates —absorbed by plants→ Plant protein —eaten by animals→ Animal protein

During an animal's lifetime, it will produce waste in the form of urine and faeces. These will be acted upon by the decomposers and gradually broken down. This results in nitrates and other nutrients being released into the soil.

When a plant or animal dies, the same process of decomposition takes place.

The breakdown of the proteins in the waste material and in the bodies of dead organisms takes place in stages. Each stage is caused by a different group of decomposers:

- The proteins are broken down to ammonium compounds by fungi and bacteria.
- The ammonium compounds are changed to nitrites by nitrifying bacteria.
- The nitrites are finally changed to nitrates by nitrifying bacteria. Nitrates are therefore available again for new plants to absorb.

The complete pathway that nitrogen takes in the ecosystem is called the nitrogen cycle.

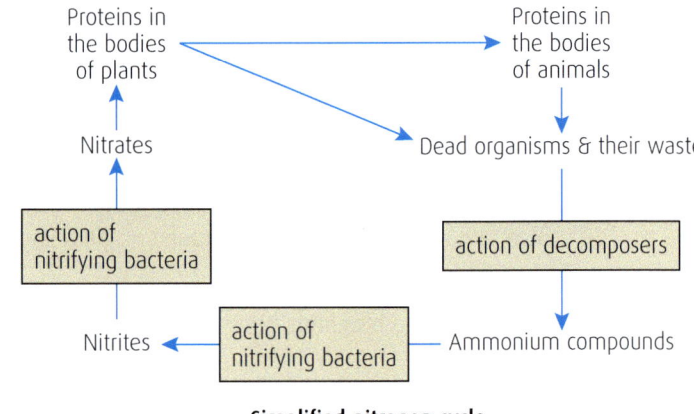

Simplified nitrogen cycle

The Nitrogen Cycle – Recycling U3

MORE ABOUT THE NITROGEN CYCLE

- **Root nodules** – Some plants have an advantage over others in terms of gaining nitrogen. These plants benefit by having little lumps on their roots, known as root nodules. The nodules contain bacteria, called nitrogen-fixing bacteria, which convert nitrogen from the air into nitrates. This means that the plant has its own supply of nitrates for making protein. Such plants are known as leguminous plants. Examples are peas, beans and clover. Similar bacteria are also free-living in the soil, producing soil nitrates from atmospheric nitrogen.
- **Lightning** – When lightning occurs, a small quantity of nitrogen in the air is converted into soil nitrates which become available for plants to absorb.
- **Denitrifying bacteria** – These bacteria use nitrates in the soil and release nitrogen gas into the atmosphere.

Root nodules

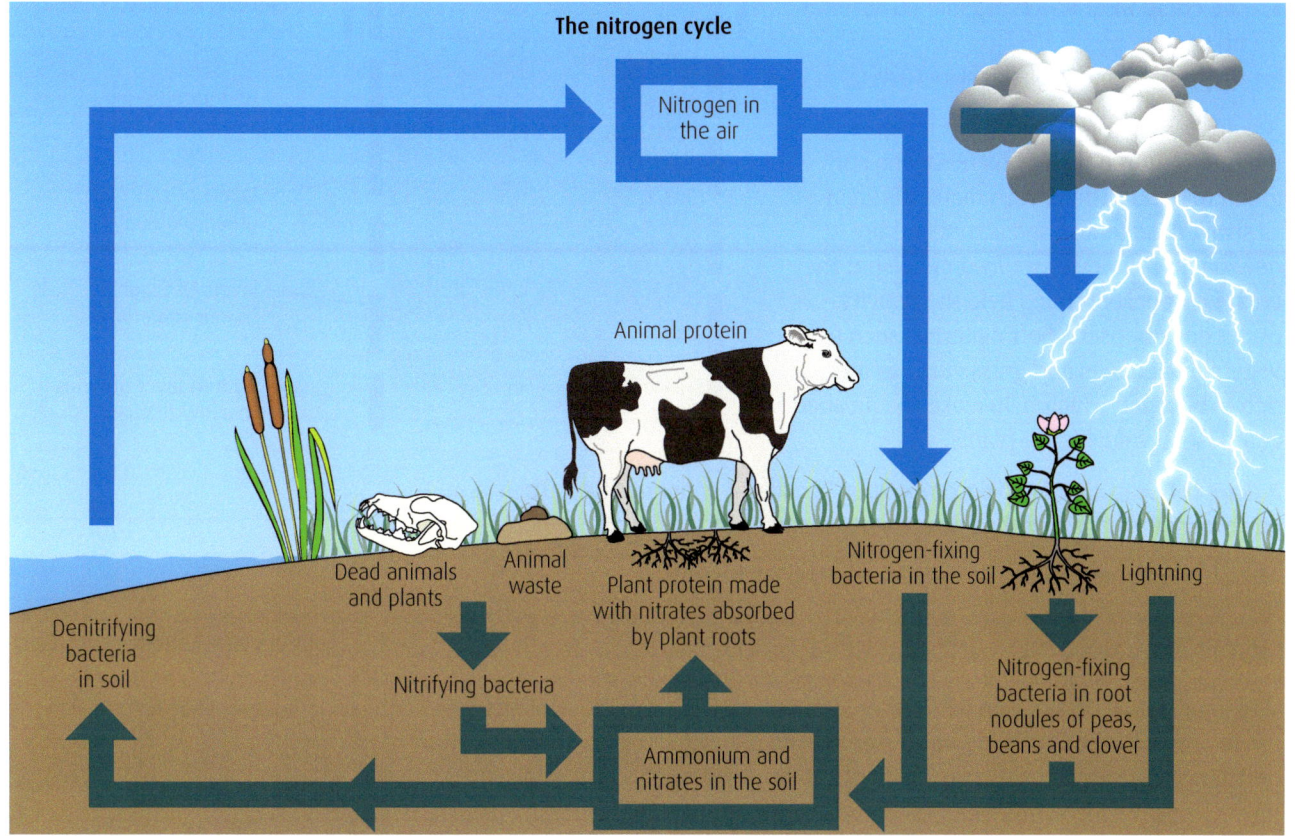

SUMMARY

The nitrogen cycle is a series of processes in which nitrogen circulates in the environment. It includes proteins in the bodies of organisms, nitrates and other chemicals in the soil and atmospheric nitrogen. Various bacteria and fungi are important for some of the stages of the cycle.

DON'T FORGET

Decomposers are fungi and bacteria which break down waste materials and dead organisms to produce nitrates that living plants can absorb.

THINGS TO DO AND THINK ABOUT

1. Name two types of organisms which are decomposers.
2. What use do plants make of absorbed nitrates?
3. Why do plants with root nodules have an advantage over other types of plants?
4. Nitrogen is lost from the soil when plants absorb nitrates. Describe one other way in which nitrogen can be lost from soil.

THE NITROGEN CYCLE

COMPOST HEAPS & WATER CULTURE EXPERIMENTS

COMPOST HEAPS

A compost heap is a method of speeding up the natural decomposition of dead plant material to recycle valuable plant nutrients back into the soil. The advantages of composting include:

- the reduction of waste material going to landfill sites
- the creation of a rich organic fertilizer at no cost
- the improvement of soil structure.

A compost heap works best if it includes a mixture of tough fibrous material and soft green material. The fibrous materials such as straw, dead leaves and scrunched up newspaper are slower to decompose but they create air pockets which help the activity of the decomposers. Soft materials such as vegetable peelings and grass cuttings are rich in nitrogen which help make the finished compost so valuable for plants.

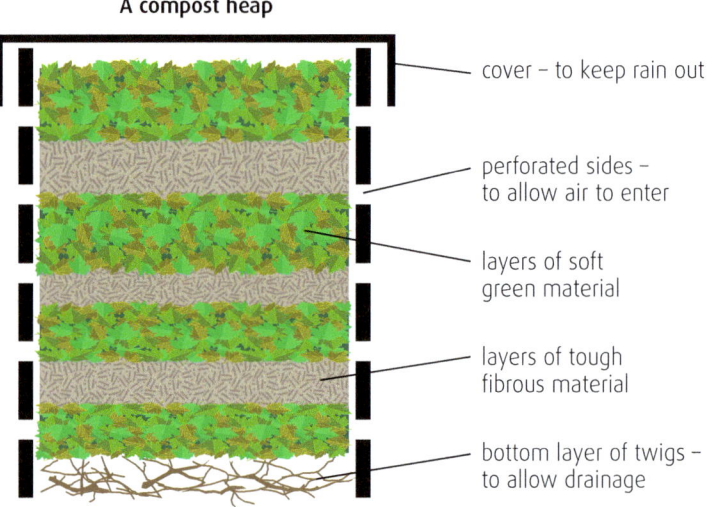

A compost heap
- cover – to keep rain out
- perforated sides – to allow air to enter
- layers of soft green material
- layers of tough fibrous material
- bottom layer of twigs – to allow drainage

A compost heap speeds up decomposition because it generates heat. The warm environment means that the microbes which carry out the decomposition are more active.

The finished compost can be spread over soil or dug into it.

WATER CULTURE EXPERIMENTS

The importance of nitrogen (and other elements) to the growth of plants can be demonstrated by carrying out water culture experiments. Plants suffer from deficiency diseases if particular nutrients are in short supply. These diseases have characteristic symptoms which can be observed in the plant (see page 61).

The following investigation can be carried out to demonstrate the effect of nitrogen deficiency.

- Two identical containers are each fitted with a polythene bag lining.
- Pure (distilled) water is added to each bag.
- One container should then have all the minerals needed for healthy plant growth added to the water, while the other container has the same minerals except for nitrogen.
- Seedlings of the same variety are floated on the surface in each container and they are left in the same conditions for the plants to grow.
- After a period of time, the plants are examined and differences between the growths of those in each container noted.

Plants which suffer from a deficiency of nitrogen grow very slowly as they do not make enough protein. They have yellowish leaves and this affects the process of photosynthesis. If they are grown as crop plants, then the yield of that crop will be very low. Sometimes the bases of the leaves have a reddish colour.

Nitrogen requirement of plants

All minerals present

Lacking nitrogen

The Nitrogen Cycle – Compost Heaps & Water Culture Experiments U3

Nitrogen requirement of plants

Increasing nitrogen concentration

It is therefore important for farmers to make sure that the soil in which they grow their crops has an adequate supply of nitrates in it. This will allow the growth of healthy crops and give a good yield. However, when the crop is harvested, the nitrogen is taken out of the cycle and the field may not have enough nitrates to support another crop. Therefore more nitrate needs to be added to the soil. This can be achieved by adding fertiliser to the soil.

DON'T FORGET

Nitrogen in the form of nitrates is needed by plants to produce proteins. Animals obtain the nitrogen they require by eating plant protein.
Excess nitrogen fertiliser can cause environmental problems (see pages 50 and 72).

SUMMARY

- Compost heaps help recycle waste plant material. They reduce the amount of waste going to landfill sites and create nutrient-rich compost which improves soil and helps plant growth.
- Water culture experiments can show deficiency symptoms associated with the lack of individual plant nutrients.

THINGS TO DO AND THINK ABOUT

1. Why does decomposition happen faster in a compost heap than on the soil surface?
2. Tough fibrous materials take a long time to decompose. Why is it important that they are included in a compost heap?
3. What symptoms are shown by plants which lack nitrogen?
4. Why is it important that farmers ensure their soil has plenty of nitrates in it?

FERTILISER DESIGN AND ENVIRONMENTAL IMPACT OF FERTILISERS

MINERAL DEFICIENCIES & FERTILISER DESIGN

MINERAL DEFICIENCIES

All plants require minerals to grow properly. Each mineral is only needed in very small quantities but, if any one of them is missing or in short supply, plants will not show healthy growth. For farmers and other plant growers it is essential that they ensure the soil their plants are growing in has an adequate supply of all the essential minerals.

As discussed in the previous section, plants require nitrogen to make proteins. If the soil has insufficient nitrogen in it in the form of nitrates, plants will suffer from a deficiency disease. There are other minerals which plants require for healthy growth. The two other main minerals are phosphorus and potassium. Again, if the soil lacks enough of these, the plants grown there will suffer from deficiency diseases.

Usually these three minerals are referred to by their chemical symbols, rather than having to write out their full names.

Role of NPK in plants

Mineral	Chemical symbol	Role in Plant
Nitrogen	N	For healthy leaves and stems
Phosphorus	P	For strong roots
Potassium	K	For formation of fruits and flowers

Deficiency diseases for each of these three important minerals can be recognised by the appearance of the plant leaves.

If a plant is lacking in nitrogen (N), it will have very pale green or yellow leaves.

A plant lacking phosphorus (P) has reddish or purple leaf edges or bases.

A plant lacking potassium (K) has yellowish leaves and can have purple spots. (See page 61 for more information).

Plant growers regularly check samples of their soil to make sure that there is an adequate supply of these necessary minerals. If they find that there is a shortage of one or more of the minerals, they will take steps to remedy this before planting their crops.

Harvesting crops removes plants and the minerals that they have absorbed from the soil, so farmers have to replace these minerals by adding fertilisers to the soil. Fertiliser is a term applied to any material that is used to supply the minerals required by plants to allow healthy growth.

Potassium
for
Flower head
or Fruit

Nitrate
for
Leaf and
Stem

Phosphate
for Root
System

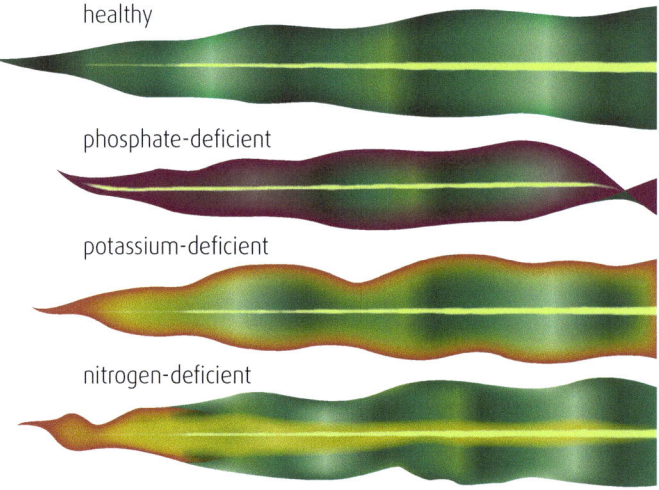

healthy

phosphate-deficient

potassium-deficient

nitrogen-deficient

Deficiency diseases in plants

Nitrogen (N) is needed for healthy leaves and stems, phosphorus (P) for strong roots and potassium (K) for the formation of flowers and fruits.

FERTILISER DESIGN

There are different methods of improving the fertility of soils to make sure that the plants growing in them have a sufficient supply of the essential minerals. Fertilisers increase the crop yield and some can also help to improve soil texture.

Fertilisers can be natural or artificial (man-made). They come in different forms, for example, liquids and pellets, and can be applied in different ways, for example, sprayed on or ploughed in.

NATURAL FERTILISERS

Green manure growing

- **Green manure** – Sometimes when a field has been used for a crop that has been harvested, a farmer may decide not to grow another commercial crop in the field. Instead a plant such as clover is grown and then it is ploughed back into the soil. The reason for this is that clover has root nodules which contain bacteria which convert nitrogen from the air into nitrates. By ploughing the clover back into the soil, the nitrates are released for other plants to use. This therefore increases the 'nitrogen value' of the soil. Plants such as clover are sometimes known as 'green manure'.
- **Animal manure** – Another way to increase the nitrogen value of the soil is to use animal waste material (urine and faeces). Farmers collect the waste from their animals and mix it with water to form liquid manure called slurry. This is then sprayed onto the soil. Microorganisms (bacteria and fungi) break down the waste, eventually producing nitrates for plants to absorb.
- **Artificial fertilisers** – These are chemicals which are manufactured. Most are produced by a reaction between an acid and an alkali. The neutral solution produced is evaporated until crystals form. These are filtered out of the mixture and dried. Artificial fertilisers can be applied as a solid or a liquid. Most are applied as solid fertilisers which dissolve when it rains and the liquid produced seeps into the soil, providing minerals for plant growth. Liquid fertilisers are usually concentrates and are diluted with water before use.

The benefits of green manure

SUMMARY

Plants need to absorb small amounts of minerals from the soil to grow properly. The most important minerals are nitrogen, phosphorus and potassium. If plants do not have enough of a mineral, they will show an associated deficiency disease.

Plant growers add fertilisers to the soil to ensure an adequate supply of essential minerals.

Fertilisers can be natural or artificial.

Artificial fertiliser

THINGS TO DO AND THINK ABOUT

1. Name each of the three main minerals needed for healthy plant growth. For each, give its chemical symbol and why a plant requires it.
2. Why is harvesting a crop from a field likely to lead to mineral deficiency in that field?
3. Name two types of natural fertilisers.
4. Explain why growing plants with root nodules in a field benefits later crops grown in that field.

FERTILISER DESIGN AND ENVIRONMENTAL IMPACT OF FERTILISERS

DIFFERENT FERTILISER TYPES & EUTROPHICATION

ADVANTAGES AND DISADVANTAGES OF DIFFERENT FERTILISER TYPES

NATURAL FERTILISERS

- **Advantages** – Using natural fertilisers is the least expensive method of adding fertiliser. It adds organic material, improving the texture as well as the nutrient levels of the soil. The addition of organic material also increases the ability of the soil to hold water, reduces the chance of erosion by water and wind and raises the pH of the soil. It can be made on a large scale on farms or domestically from kitchen scraps and grass clippings which are left to rot on compost heaps.

  Natural fertilisers are less likely to damage young plants as they are less concentrated than artificial chemical-based fertilisers.

- **Disadvantages** – Natural fertilisers are slow to break down and are often quite smelly and unpleasant to apply. As the various organic materials break down at different rates, the level of nutrients in this type of fertiliser is not consistent.

The NPK ratio can be seen on the front of packets of chemical fertilisers

ARTIFICIAL CHEMICAL FERTILISERS

- **Advantages** – These can be manufactured to provide specific levels of nutrients. This means that they are predictable and reliable in their content. Different 'blends' can be tailor-made to suit particular types of plants. The quantity of the three main ingredients (nitrogen, phosphorus and potassium) can be carefully controlled in each mix. This is usually written on the packet as an N-P-K ratio. This allows the person applying the fertiliser to know exactly which nutrients are present and how much of them are being applied to wthe soil.

  JUST A WEE NOTE
  Phosphorous is often referred to as phosphate in chemical fertilisers.

- **Disadvantages** – Commercial chemical fertilisers are more expensive than natural types. Care is needed in handling them as they can cause skin problems and affect the respiratory system if breathed in. It is important to measure out the dosage accurately as too much can damage or even kill plants. Using too much can also mean that the chemicals build up in the soil, affecting the soil pH and fertility.

Fertiliser burn on a leaf

DON'T FORGET

Bulky fertilisers such as manures can help to improve the structure of the soil, as well as providing nutrients for plants.

Fertiliser Design and Environmental Impact of Fertilisers – Different Fertiliser Types & Eutrophication

EUTROPHICATION

Unfortunately soluble fertilisers can be washed away by rainwater into rivers and lakes. This causes an increase in the level of nutrients in these water systems. The result of this is that the algae in the water grow much faster than normal. This process is called 'eutrophication'.

As the algae increase rapidly they form a carpet over the surface of the water, blocking out the light. This is known as an algal bloom.

Other plants growing in the water die off due to lack of light for photosynthesis and eventually some of the algae die off too. All these dead plants provide food for bacteria, which greatly increase in number. The vast number of bacteria use up so much oxygen for respiration that there is not enough left to support all of the other living things. In extreme situations, all life in the river or lake can die.

Process of eutrophication

normal lake

lake during eutrophication

Result of eutrophication

SUMMARY

Natural fertilisers include green manure, animal manure and compost. They are inexpensive, bulky and can improve soil structure. They are less consistent than artificial fertilisers.

Artificial chemical fertilisers can be manufactured with different concentrations of the essential nutrients to suit different conditions. Applied in the correct way and in appropriate quantities, fertiliser will help plants to grow better; too much fertiliser can however be damaging to plants and animals.

THINGS TO DO AND THINK ABOUT

1. Give an example of a natural fertiliser and explain an advantage and a disadvantage of using it.

2. What does it mean if a fertiliser packet has 30-10-10 written on it? Which part of a plant will it encourage most to grow?

3. Explain each of the following points about eutrophication.

    (a) The source of the excess nutrients in the water.

    (b) The type of organism which forms a 'bloom' on the water.

    (c) The reason for the increase in number of bacteria in the water.

    (d) The reason why the increase in bacteria leads to the death of other aquatic organisms.

ADAPTATIONS FOR SURVIVAL

THE NEED FOR AND THE TYPES OF ADAPTATION

THE NEED FOR ADAPTATION

In order for plants and animals to survive in particular environments, they must be adapted to the conditions found there. An adaptation which allows an organism to survive in one area might mean that it cannot survive elsewhere. For example, a polar bear is adapted to live in the Arctic and could not survive in the desert.

Adaptation is an evolutionary process, allowing a species to be better able to live in its habitat. Adaptations usually arise because a gene mutates by accident. If the mutation helps an individual plant or animal to survive better than others which do not have the mutation, it will get passed on to the next generation. The enormous variation shown, even between members of the same species, helps in their survival.

For example, imagine a species of mouse. If a mouse should be born with longer legs than all the other mice in the species, it will be able to run faster and escape predators better than all the others. It is therefore able to live longer and breed, passing on the gene for longer legs on to its offspring. They may also live longer and have more offspring and the gene will continue to be inherited generation after generation, thus continuing the adaptation down the line.

Mouse

TYPES OF ADAPTATION

Adaptations of plants and animals can be divided into three broad categories: structural, physiological and behavioural.

STRUCTURAL ADAPTATIONS

Structural adaptations are physical features of an organism. Examples could include features such as the shape of bird beaks, shape of leaves on trees, spines instead of leaves on cactus plants and size of fish fins.

PHYSIOLOGICAL ADAPTATIONS

Physiological adaptations are changes within the cells or tissues of an organism which are made in response to an external stimulus. This change improves the ability of the organism to survive in that environment. Examples could include the ability of snakes to produce venom or the ability of camels to produce very concentrated urine and so reduce water loss.

BEHAVIOURAL ADAPTATIONS

Behavioural adaptations are actions organisms carry out in order to allow them to thrive and survive. The behaviour can be learned and passed on through the generations or it may be instinctive behaviour passed on genetically. Examples could include migration of birds to warmer climates at certain times of the year, hibernation of animals or animals being nocturnal.

Adaptations may be associated with the structure of an organism, the way its body functions or with how it acts.

> **DON'T FORGET**
> Adaptations arise as a result of mutations and give rise to variation in a species.

> **DON'T FORGET**
> Adaptations are necessary if organisms are to survive and thrive in changing environments.

Adaptations for Survival – The Need for and the Types of Adaptation

ADAPTATIONS OF PREDATORS AND PREY

PREDATORS

Successful predators are those which catch plenty of prey. In order to do so they are adapted in several ways, including features such as:

- streamlined, muscular bodies to allow them to run at speed
- sharp claws and large pointed teeth to catch hold of prey
- eyes which both face the front, giving them the sort of vision which allows the judgement of distance.

Foxes are examples of successfully adapted predators

PREY

Animals which are preyed upon need to be adapted to avoid predators. Such adaptations could include:

- streamlined, muscular bodies to allow them to run at speed
- camouflaged coats to avoid being seen
- social behaviour where they live in herds/flocks or other groups
- defence mechanisms, for example, the sting of a wasp
- eyes situated at either side of their heads, allowing them to see all around them.

An example of an animal which has adapted to avoid predators is the rabbit

ADAPTATIONS FOR POLLINATION IN FLOWERING PLANTS

Plants may show adaptations which help the method of pollination they use for reproduction. Pollination can either be carried out by insects or by the wind. The table shows adaptations of flowers pollinated by each of these methods.

Feature	Insect pollination	Wind pollination
Pollen	Sticky pollen	Small, lightweight pollen
Petals	Colourful petals	Dull petals
Nectar	Sweet, sugary nectar	No nectar
Stigma	Sticky stigma	Feathery stigma
Scent	Sweet scent	No scent

These adaptations ensure that at least some of the pollen produced by the flowers is picked up and carried either by insects or wind to the stigma of another flower, enabling the process of reproduction to take place.

Insect-pollinated flower

Wind-pollinated flower

> **SUMMARY**
> - Adaptations are features of an organism which helps it to survive by making it better suited to its environment. Adaptations are normally the result of beneficial mutations.
> - They may be structural, physiological or behavioural.

THINGS TO DO AND THINK ABOUT

1. Why is it necessary for organisms to adapt to their environment?
2. Name the three different types of adaptations found in living organisms and give an example of each.
3. Explain why many predators have both eyes facing forward while prey often have eyes at the sides of their heads.
4. (a) List three differences between flowers that are pollinated by insects and flowers that are wind pollinated.
   (b) For each of the flower features you have listed, explain why it is adapted in this way.

ADAPTATIONS FOR SURVIVAL

EXAMPLES OF ADAPTATIONS

POLAR BEARS

These animals have a number of adaptations which allow them to survive in the extreme conditions found in the Arctic:

- White fur allows them to blend in with the snow and ice of their surroundings, giving them camouflage when hunting.
- A thick layer of fat under their skin helps them to keep warm. Without it they would lose heat quickly and could die.
- Fur on the soles of their feet provides insulation and gives them better grip on the snow and ice.
- Large wide paws spread their weight, helping to prevent them falling through the snow. Large feet also help them to swim well.
- Polar bears are so well adapted to their cold environment that sometimes they can become overheated. If this happens they will swim to cool down.

CAMELS

Deserts are areas which typically have very little water and high daytime temperatures. The adaptations of camels which aid their survival include:

- Large feet to spread their weight, allowing them to walk on sand without sinking into it.
- Long bushy eyelashes and the ability to close their hair-lined nostrils help to prevent sand from getting into their eyes and nose.
- One or two humps containing fat as an energy store. Having fat contained in one area means it does not insulate the body. This prevents overheating.
- Tissues which can tolerate water loss much more than other animals, without the cells dying.
- Tissues which can tolerate high temperatures. This means that they do not need to sweat to cool down. This conserves water.

Bactrian camel

Dromedary

CACTUS PLANTS

Cactus plants also have several adaptations for survival in the hot, dry climate of deserts.

- Their rounded shape loses less water than a thin flat shape would.
- They have a thick waxy cuticle which is waterproof, therefore helping to conserve water in the plant.
- Their thick stems can store water.
- They have reduced their leaves to narrow spines which reduces water loss through leaf surfaces. This adaptation also helps to prevent them from being eaten by animals.
- They have green stems which allow the stems to photosynthesise, which is important as the leaves are reduced to spines.
- Their widespread roots allow them to collect water over a large area when it is available.

Adaptations for Survival – Examples of Adaptations

FISH

Rivers, lochs and oceans provide many different habitats. Each habitat contains species of fish adapted to survive in the conditions found there.

Most fish that are adapted to survive in fresh water could not tolerate the conditions in salty sea water and vice versa.

However, there are a few species such as salmon and eels which spend part of their lives in rivers and part in the sea. Scientists are still investigating the physiological adaptations of these fish to try to understand how they are adapted to survive in two such contrasting environments.

Examples of fish with specific adaptations:

ONLINE

Some websites which may help with Questions 3 and 4 are:

http://www.atlanticsalmontrust.org/salmon-life-cycle-habitat-threats-and-concerns.html

http://www.unm.edu/~toolson/salmon_osmoregulation.html

http://www.nhs.uk/Livewell/bites-and-stings

http://mic-ro.com/plants

http://www.thepoisongarden.co.uk/

1. STURGEON

Sturgeons are marine fish. They are the source of the delicacy, caviar. Their adaptations include:
- Sucker-shaped mouths under their heads suck up food like a vacuum.
- Whisker-like structures called barbels near their mouth detect food by feeling it on the sea bottom.
- Small eyes, since large eyes are not needed for finding food in the dark waters on the sea bottom.
- A dark-coloured back and a light coloured belly make them less visible to predators both above and below them.

2. SEAHORSES

The adaptations of seahorses include:
- Eyes on either side of their head which can move independently of each other allows them to detect predators.
- Tails able to grasp onto things allow them to anchor in one place, rather than being drawn away by water currents.
- A layer of mucus covering their skin protects them from the stings of sea anemones where they hide for protection.
- A variety of skin patterns and colours provide camouflage in their environment.

3. PUFFER FISH

Puffer fish have several adaptations to aid their survival. These include:
- The ability to take in vast quantities of water (or sometimes air) into their very elastic stomachs makes them inflate into a ball-like shape, preventing predators swallowing them.
- Spines in the skin of some species make them less appealing to predators.
- Production of a toxin which is 1200 times more poisonous that cyanide makes them foul tasting and lethal to predators.

SUMMARY

Every habitat has its own characteristic conditions. Adaptations allow organisms to cope with the conditions of their particular habitats. The better adapted they are, the greater their chances of survival.

DON'T FORGET

An adaptation is a characteristic, usually inherited, which increases an organism's chances of survival in its habitat.

THINGS TO DO AND THINK ABOUT

1. From the list of adaptations given about polar bears, identify one which is a structural adaptation and one which is a behavioural adaptation.
2. From the list of adaptations of camels, which are physiological adaptations?
3. Find out which part of a salmon's life is spent at sea and when it lives in a river. Try to discover why it is a problem for the fish to live in these two very different environments.
4. Some plants make chemicals to sting or poison animals which might eat them. Find the name of a plant which stings and one which makes a poison.

LEARNED BEHAVIOUR IN RESPONSE TO STIMULI

INNATE BEHAVIOUR

Animal behaviour refers to the way in which animals respond to the stimuli they encounter. Behaviour can be innate (instinctive or inborn) or it might be learned behaviour. In both cases, appropriate behaviour increases the chances of an animal's survival.

Innate behaviour is inherited genetically and does not need to be learned.

SWARMING

Some animals gather in large groups and often travel to new locations together. Examples of this are seen in birds such as starlings and also in bees when they move to a new hive. Staying close together in large numbers decreases the chances of any individual being killed by a predator.

Starlings swarming in the sky

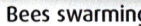
Bees swarming

HUDDLING

Some animals, for example Emperor penguins, huddle together to keep warm. This allows them to survive very harsh weather conditions which would possibly be fatal to an individual left on its own.

Emperor penguins huddle to stay warm

MIGRATION

Many animals migrate at specific times of the year. They can travel very long distances in order to find warmer climates, better food availability and places to raise young. These migrations often take place annually.

Examples include:

- Wildebeest which migrate over 1800 miles each year in search of new grassland.
- Swallows which migrate from Britain to Africa to escape our winters and from Africa to Britain in spring to breed.
- Salmon which migrate to the sea for several years and then migrate back to the river they were born in to breed.
- Emperor penguins which migrate over 60 miles through Antarctica to breed.

Wildebeest migration

Salmon returning to the river to spawn

TERRITORIAL BEHAVIOUR

Some animals establish their own territory and defend it from other members of their species. The area an individual defends must be large enough to provide sufficient food for itself, its mate and its offspring. The area must not be so large that the animal has problems defending it. Such behaviour is territorial. Animals will often mark out the area that they establish as their territory and warn others off if they stray into it.

An example can be seen in the garden robin. The male bird will fiercely defend his territory against other robins. He puffs out his red breast and sings a high-pitched song to warn off others. He becomes most aggressive to intruders if they are at the centre of his territory and may attack them.

Advantages of this type of behaviour are that it ensures enough food to raise young as well as providing them with a safe and protected environment.

> **DON'T FORGET**
>
> Innate behaviour is beneficial because an animal is born already able to make some responses which may aid survival.

SUMMARY

- Animals behave in certain ways in order to increase their survival chances.
- Behaviour can be innate (instinctive) or it can be learned.
- Swarming, huddling, migration and territorial behaviour are examples of innate behaviour.
- Territorial behaviour means that an animal defends an area where they can find shelter and sufficient food to share with a mate and raise a family.

THINGS TO DO AND THINK ABOUT

1. What term is used to describe behaviour that is inborn and does not have to be learned?
2. What type of behaviour is shown by millions of locusts moving together in a large group?
3. Some birds show territorial behaviour.
   (a) What is the advantage to a male robin of establishing his own territory?
   (b) Describe how a male robin behaves if another robin enters its territory.
4. Swallows migrate to Britain in spring to breed because there are more hours of daylight. Suggest why this is an advantage for raising young.

LEARNED BEHAVIOUR IN RESPONSE TO STIMULI

LEARNED BEHAVIOUR

Learned behaviour is not inherited but is gained by experience or by watching others.

GROUP BEHAVIOUR

Japanese macaques live in cold areas of Japan, in troops of 20–30 individuals.

They learn from one another, allowing new, beneficial behaviour to be passed from one individual to another. Many years ago, individual monkeys learned that hot springs in their environment were a source of warmth. Nowadays whole troops can be found bathing in the springs.

Macaque monkeys keeping warm

IMPRINTING

Some young animals learn to follow the first large moving object they see. This is usually their mother. The young benefit because their mother will provide food and protection for them.

Ducklings imprinting on their mother

DON'T FORGET

Behavioural responses of animals increase their chances of survival.

COMMUNICATION

Communication between individuals allows knowledge to be passed on, for example, the waggle dance of bees. Bees forage for nectar and pollen which they bring back to the hive. A bee, on finding a good food source, returns to the hive and performs a waggle dance. This indicates the distance and direction of the food source to other bees. This can save them wasting time and energy searching on their own. Bees are good learners. A bee might fly to several different types of flowers in a morning. If one type has a good supply of nectar and pollen, it will continue to visit that particular kind of flower for the rest of the day. It is thought that it recognises the flower type by its colour and its scent.

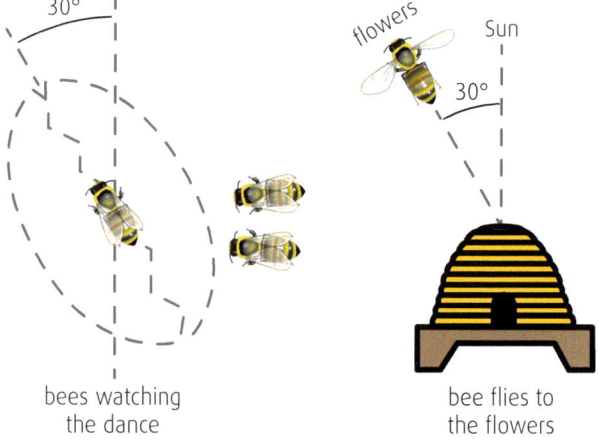

Explanation of waggle dance

HABITUATION

Many animals show avoidance behaviour when danger threatens, for example, the withdrawal reflex of snails. If the antenna of a snail is touched it withdraws into its shell. This is an innate response. However, it can learn to ignore the stimulus if it is harmless. This avoids wasting time and energy retreating when there is no need. It can be easily demonstrated:

- Place a large snail on a clean surface and wait until it has fully emerged from its shell.
- Touch one of the antennae with a damp cotton bud.
- The snail retreats into its shell.
- Repeat the procedure 10 times.

Eventually the snail learns to ignore the harmless touch. If the experiment is repeated at a later time, the snail will again withdraw into its shell, showing that the learning is very short-lived.

DON'T FORGET

Learning to behave in a certain way in response to stimuli in the environment can help to increase survival chances.

Learned Behaviour in Response to Stimuli – Learned Behaviour U3

METHODS OF RESEARCHING AND DEMONSTRATING ANIMAL BEHAVIOUR

CHOICE CHAMBERS

Choice chambers are used to investigate the behaviour of small invertebrates such as woodlice. The choice chamber is a shallow dish containing two contrasting 'environments', one in each half, for example moist and dry or dark and light.

Several woodlice are placed into the dish. After a suitable period, the number of woodlice in each side is counted. The experiment should be repeated several times and average results calculated. The area in which most woodlice are found indicates which of the conditions is more suitable for their survival.

A choice chamber with moist and dry sides.

MAZES

A maze can be used to investigate the ability of animals to learn. A simple T-maze provides an animal with a choice of turning left or right. A food reward is placed at Side A or B. A hungry rat enters the maze and turns left or right at random. If it always finds food at one side, can it learn always to turn in that direction? Things to investigate include:
- How many times does the experiment need to be repeated for the rat to always choose the correct arm?
- What happens if the food is then removed?
- If the food is then replaced, does the rat go back to its original behaviour?

Simple T-shaped maze

LEARNING IN HUMANS

Simple investigations can demonstrate learning in humans. If an unfamiliar task is given to a subject then errors are likely to be made. If the task is repeated a number of times, then the subject becomes better and fewer mistakes are made. The number of errors made at each attempt is recorded and plotted on a graph. Such graphs are called 'learning curves'.

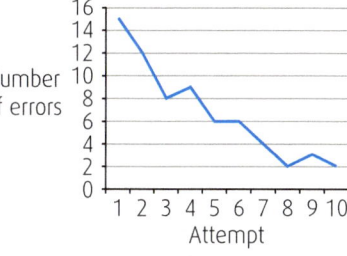

Mirror drawing learning curve

Examples of such tasks are:
- Mirror drawing: this involves tracing over a complicated pattern by only looking at the image in a mirror. An error is recorded each time the subjects pen moves off the pattern.
- Typing word lists: this involves typing a long list of simple words in a set time. The results can be recorded as the number of errors made or as the number of words typed correctly.

SUMMARY
- Sometimes an animal will carry out a chance action which is of survival value.
- This behaviour can be copied by other individuals, increasing their survival chances.
- Simple investigations can be carried out to show that animals can learn to respond in certain ways to increase their chances of finding food or environments that are suited to them.
- Individuals can learn to ignore repeated stimuli if they prove not to be a threat, but this type of learning is short-lived.
- Humans can learn by practising a task repeatedly.

THINGS TO DO AND THINK ABOUT

1. Explain fully why it is good practice to repeat an experiment several times and calculate an average.
2. Sometimes an animal learns to ignore a stimulus which it originally treated as a threat. Why is it important that, if the stimulus occurs a while later, the animal once again reacts to it as a threat?
3. In humans, learning can take place by continually repeating a task. Why does the time to complete a task get faster each attempt?
4. Why do animals such as bees use a communication system to tell others where food is rather than leaving them to find it for themselves?

COURSE ASSESSMENT

AN OVERVIEW OF THE ASSESSMENT

The good news is that you are not required to sit a final exam at the end of your National 4 Biology studies in order to achieve a course award. Instead of a final exam you are required to pass four assessment tasks. These four tasks will take place in school and they will be marked by your teacher on a pass or fail basis.

WHAT ARE THE ASSESSMENT TASKS?

Assessment task 1   Preparation of a scientific report on a biology experiment or practical investigation.

Assessment task 2   A short scientific report based on research of a biology topic.

Assessment task 3   A set of questions on the biology you have covered. This will cover Knowledge & Understanding as well as Problem-Solving Skills and will be completed under exam-style conditions.

Assessment task 4   This task is called the Added Value Unit or Assignment. This will be based on a topical issue relating to your National 4 course and its effect on society or the environment.

ASSESSMENT TASK ONE

To be successful in this assessment you must plan and carry out an experiment or practical investigation and then prepare a scientific report about it.

Your report must cover the following areas:

1. PLANNING

This section of the report must include:

(i) The aim of the experiment, that is, what you are trying to find out. Examples of suitable aims could be:
   a. To find out how the concentration of fertiliser applied affects the growth of plants.
   b. To investigate the effect of pH on the activity of the enzyme amylase.

(ii) The variable(s) that need to be kept constant.

(iii) Measurements or observations to be made.

(iv) The resources (apparatus and materials) to be used.

(v) The procedure, that is, how the experiment will be carried out. This must be written in such a way that another National 4 pupil could use the information to repeat the experiment.

(vi) Any safety precautions needed and the reasons for them. This could include the use of safety glasses if a chemical is an irritant which could damage eyesight or not using naked flames while using flammable liquids as they could easily go on fire.

2. FOLLOWING PROCEDURES SAFELY

Your teacher will check that you carry out the experiment in a safe manner.

3. MAKING AND RECORDING OBSERVATIONS/MEASUREMENTS ACCURATELY

4. PRESENT RESULTS IN AN APPROPRIATE MANNER

Some things you might include are:

(i) Tables with appropriate headings and units.

(ii) Calculations of average results from repeated experiments.

(iii) Flow charts.

(iv) Bar charts or line graphs. Charts or graphs should be drawn on graph paper and have appropriate scales, labels and units. The bars or points should be plotted correctly.

5. DRAWING VALID CONCLUSIONS

Your conclusion must relate to your aim.

The valid conclusions for the aims given in Section 1, Planning, might be written as

a. The greater the concentration of fertiliser used, the greater the growth of the plant.
b. As the pH increased from 1 to 7, the activity of amylase increased.

6. EVALUATING THE PROCEDURE

In this final section of the report you must suggest something that would improve the experiment.

> **DON'T FORGET**
>
> A variable in an experiment is something which can be altered. For example, the volumes of solutions used, the concentrations of solutions used, the temperature and the pH used are all variables. For an experiment to be valid only one variable should be changed. Other variables must remain unchanged.

Course Assessment – An Overview of the Assessment

ASSESSMENT TASK TWO

To be successful in this assessment you must select and research an application of biology from one of the key areas you have studied. The report should show the impact of the application on society or the environment.

When you have completed your report you should read over it and make sure that you have:

(i) Stated the application of biology involved.

(ii) Used appropriate biology knowledge to describe the application. For example, if your application was the use of enzymes in industrial processes, you could give information about enzymes speeding up chemical reactions without being used up in the process.

(iii) Stated how society or the environment has been affected by the application.

(iv) Used appropriate biology knowledge to describe its effect. For example, if your application was the use of fertilisers in farming, you could state that fertilisers have a negative impact on the environment because they are soluble in water and can therefore be washed out of the soil by rain, ending up in rivers causing pollution.

DON'T FORGET
Remember to ask your teacher for the candidate guides for the assessment tasks.

ASSESSMENT TASK THREE

This is a set of questions on the biology you have studied. It will consist of knowledge based questions (assessment standard 2.1) and problem solving questions (assessment standard 2.4). There will be three types of problem solving; predicting, selecting and processing.

Below is an example of each type of question.

EXAMPLE Sample question 1

A piece of onion skin was examined using a microscope and the lengths of 50 cells were measured.

The bar chart below shows the number of cells of different lengths which were found.

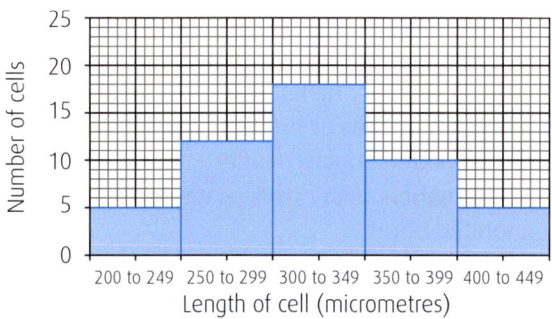

a. Which range of cell lengths contained the most cells?
   From _____
   to _____ micrometres   **1 – SELECT**

b. What percentage of cells had a length of 350 micrometres or more?
   Space for calculation

   _____ %   **1 – PROCESS**

EXAMPLE Sample question 2

The following chart gives information about cholesterol in the blood.

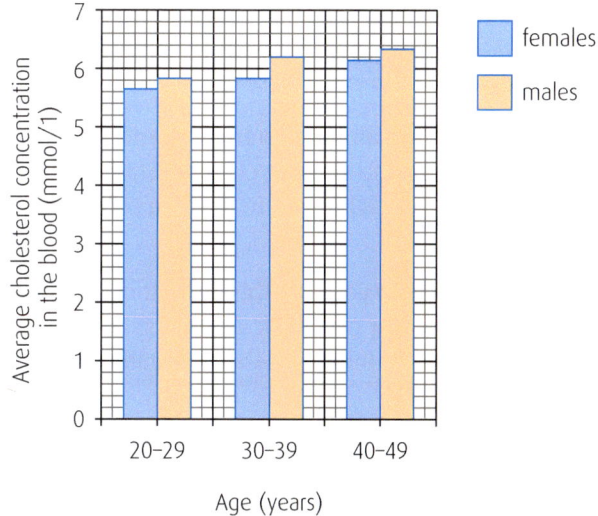

a. What is the highest cholesterol level shown for females?
   _____
                                    1 – SELECT

b. Predict the average cholesterol concentration of females aged 50–59 years if males of that age had an average concentration of 6·8 mmol/l.

   _____ mmol/l   **1 – PREDICT**

COURSE ASSESSMENT

AN OVERVIEW OF THE ASSESSMENT (contd.)

ADDED VALUE UNIT – ASSIGNMENT

The assignment at National 4 is the fourth unit and is marked as a pass or fail.

It covers the following assessment standards:

1.1 choosing, with justification, a relevant issue in biology
1.2 researching the issue
1.3 presenting appropriate information/data
1.4 explaining the impact, in terms of the biology involved
1.5 communicating the findings of the investigation.

If you fail to pass one or more of these 5 assessment standards don't panic; you are allowed one resit.

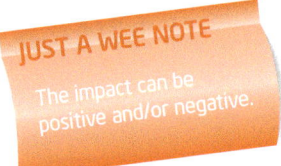

JUST A WEE NOTE
The impact can be positive and/or negative.

Apply your skills, subject knowledge and understanding to investigate a topical issue in biology and its impact on the environment and/or society.

There are two stages to the assignment process:

STAGE ONE – THE RESEARCH STAGE

First you must select a topical issue to base your assignment on. This must relate to your course work, for example:

a. the use of gene therapy in medicine
b. the use of genetic engineering in the production of Factor VIII.

Next you will gather information from a variety of sources (e.g. books/the internet/class notes/scientific journals) and this will allow you to create a candidate log or journal.

Your teacher will help you with this stage though the research must be your own work and you should record the sources for any information that you gather.

DON'T FORGET
Remember that websites which can be altered by anyone are not reliable sources.

DON'T FORGET
Remember to include the title and aim for any experiments.

If you carry out an experiment as part of the research stage, you may wish to include this work in your assignment.

STAGE TWO – THE COMMUNICATION STAGE

In this stage you will produce your final assignment. This must be all your own work.

During the second stage of the assignment you will select, use and record at least two appropriate sources from your research which will be included in your final communication.

JUST A WEE NOTE
This will take place under controlled conditions – this means your teacher will be present.

You must select, process and present information and/or data which you gathered in Stage one relating to your chosen topical issue.

You can choose how you communicate your findings and this may include one or more of the following:

- a handwritten or word-processed report (200–400 words, excluding tables and diagrams, is the suggested length)
- a presentation, oral or digital which is accompanied by supplementary material, for example, presentation notes or presentation slides with notes.
- an information booklet/leaflet
- an information poster (this poster must include annotated notes).

THINGS TO DO AND THINK ABOUT

Be sure to make good use of this book and all of your course materials while you prepare for your National 4 assessments – you have the skills and the tools necessary to succeed and as long as you do your best you will achieve a qualification of which you can be proud. Stay relaxed and focused and, most of all, good luck!

Remember to get the answers for the questions in this book at www.brightredpublishing.co.uk